A Practical Guide to
CT SIMULATION

Editors:

Lawrence R. Coia, MD
Timothy E. Schultheiss, PhD
Gerald E. Hanks, MD
Fox Chase Cancer Center

Advanced Medical Publishing
Madison, Wisconsin
1995

Published by:
  Advanced Medical Publishing
  5331 Brody Drive, Ste. 201
  Madison, WI 53705

Copyright © 1995 by Lawrence R. Coia. All rights reserved. No part of this book may be used or reproduced in any manner whatsoever without written permission from the publisher except for brief quotations embodied in critical articles and reviews. Printed in the United States of America. For information please write to the publisher.

ISBN: 1-883526-04-3

ACKNOWLEDGEMENT

The authors wish to acknowledge the financial and technical support of Picker International in bringing this book to fruition.

A Practical Guide to

CT SIMULATION

Contents

1. Introduction 1
   Lawrence R. Coia, Timothy E. Schultheiss, and Gerald E. Hanks

PART I — Technical Aspects of CT Simulation and Treatment Planning

2. Commissioning, Acceptance Testing, & Quality Assurance of a CT Simulator 5
   Kiaran P. McGee and Indra J. Das

3. Localization & Field Design Using a CT Simulator 25
   Margie Hunt

4. The Digitally Reconstructed Radiograph 39
   Indra J. Das, Kiaran P. McGee, and Gregory E. Desobry

5. CT Simulation Hardware 51
   Eric E. Martin

6. The Treatment Planning Process 61
   Margie Hunt and Lawrence Coia

PART II — Clinical Applications of CT Simulation

7. Esophageal Cancer 71
   Lawrence R. Coia

8. Head & Neck Cancer 85
   Douglas A. Fein, Andrew Shaer, and Kiaran McGee

9. Pancreatic Cancer 103
   Özer Algan, John Hoffman, and Andrew Shaer

10. Lung Cancer 115
    Nicos Nicolaou

11. Cervical & Endometrial Cancers 135
    Rachelle M. Lanciano and Steven Bonin

12. Brain Cancer 149
    Benjamin W. Corn and Walter J. Curran Jr.

13. Prostate Cancer 163
    W. Robert Lee and Gerald E. Hanks

14. Rectal Cancer 173
    Rachelle M. Lanciano and Lawrence R. Coia

15. Breast Cancer 187
    Eileen McGarvey, MD

16. Coding & Reimbursement for Conformal Therapy...197
    Jennifer O'Malley

    Appendix I: Set Up Uncertainty — Selected References 201
    Appendix II: Treatment Planning and CT Simulation Forms and Procedures................205
    Index 207

Chapter 1

INTRODUCTION

Lawrence R. Coia, MD, Timothy E. Schultheiss, PhD,
and Gerald E. Hanks, MD

The revolution in diagnostic imaging that resulted from the introduction of the CT scanner was so profound that it was recognized by the awarding of a Nobel Prize in Physiology and Medicine in 1979. Nonetheless, there were many years between the introduction of the first commercial CT scanners and the use of modern CT scanners in radiation oncology treatment planning.

Prior to the days of the CT scanner, tumor localization was accomplished using plain films, nuclear medicine procedures, and anatomical knowledge. The imaged and expected tumor location was drawn on the simulation film and an actual field was defined taking into account the known tumor volume as well as regions suspected of containing tumor. In radiation oncology textbooks, typical fields were often described in terms of aperture sizes and shapes for each site. Rubber stamps were made for the treatment chart as a visual aide for roughly approximating the field placement. Ultimately the emphasis on the two-dimensional representation of the field placement and beam arrangement resulted in a greater reliance on guidelines appropriate for the majority of cases, but that did not account for individual variability and did not reflect a systematic thought process applied to each patient.

The spread of CT treatment planning in the 1980s had a gradual influence on the total radiation oncology treatment planning effort. Radiation oncology departments generally had to reach an agreement with diagnostic imaging departments regarding the scheduling of patients scanned for treatment planning. A limited number of slices were obtained per patient and only a small fraction of patients was scanned for planning. There were logistic difficulties in transporting the data from the CT to the treatment planning system, in correlating the data with conventional simulation images (which still had to be obtained), and in ensuring that the same patient position was maintained throughout the several steps of the process.

Over the last decade, several factors have induced the return to the important basic concept of shaping fields by considering known and suspected tumor location as opposed to the use of semi-generic field apertures. These factors are the advent of conformal therapy, the standardization of nomenclature published in the ICRU Report 50, and the arrival of CT simulation. In conformal therapy, the intention is to conform the high-dose volume to the shape of the known (and suspected) tumor volume. ICRU Report 50 has given us the language and guidelines for defining volume of known tumor, suspected microscopic spread, and marginal volumes necessary to account for set-up variations and organ motion. In addition, language is provided that describes the volumes actually treated and allows intercomparison of dose.

It is the arrival of CT simulation that has unified the CT scanning for treatment planning, and tumor localization and simulation imaging, with full use of digital communications and imaging. Having the new capabilities of CT simulation produces challenges not previously addressed by radiation oncologists, physicists, and therapists. However, simply acquiring the technology does not guarantee quality treatment. Secure immobilization, knowledge of site specific daily variations in set-up, focused attention to known and presumed tumor localization, and a methodical analysis of portal images must accompany the well thought out use of CT simulation. To utilize CT simulation to the fullest extent possible requires a change in the radiation oncology culture.

This text represents the product of two years experience at Fox Chase Cancer Center with a commercial CT simulator connected to our treatment planning system and our PACS through local area networks. Hardware selection, acceptance testing, commissioning, quality assurance, treatment planning features, and interfaces and networking factors are covered in the technical chapters. In the clinical chapters, our approach to CT simulation in all major tumor sites is described. These solutions are neither exclusive nor exhaustive, but they should accelerate the learning process for the reader.

In CT simulation we have benefited from the expertise of diagnostic radiologists and surgeons in target volume definition. We continue to benefit from a close collaboration with Picker International in describing, developing, testing, and implementing new features of CT simulation.

— Part I —

Technical Aspects of CT Simulation *and the* Treatment Planning Process

Chapter 2

COMMISSIONING, ACCEPTANCE TESTING, & QUALITY ASSURANCE OF A CT SIMULATOR

K. P. McGee, MS and I. J. Das, PhD

INTRODUCTION

If CT simulation is to deliver upon its impressive potential, these systems must meet the rigid requirements of having small tolerances as well as being reproducible and stable over the life of the system. It is with this in mind that an acceptance testing protocol, designed at our institution, is presented. Although the protocol has been tailored to a single manufacturer's system it could, with minor modifications, be used on other systems.

THE DIFFERENCE BETWEEN SIM CT AND CT SIM

Since the integration of CT data into radiotherapy treatment staging and planning, there have been several attempts to develop systems to obtain CT quality axial reconstruction data at the time of simulation. A simulator with a CT option (sim CT) is a conventional treatment simulator that has been modified to provide axial CT data. Modifications to the simulator can be as simple as digitizing the video information from the image intensifier and as complicated as replacing the image intensifier with a linear array of detectors. Fan beam projection information is obtained by taking a single line of video information from the image intensifier along the central axis of the simulator or from the linear array of detectors. A complete set of projection information is obtained by rotating the image intensifier or detectors through 360° and applying a back projection algorithm to the data set. Image quality may be poorer when compared to images obtained from conventional CT scanners but sim CTs obtain the CT data at the time of simulation with the patient in the traditional simulation position. Webb,[1] in a review of sim CT technology, described in detail the various components of these systems and has created a comprehensive list of sim CTs, their physical characteristics, and limitations.

Unlike sim CTs, CT simulators (CT sims) differ in both purpose and functionality. The purpose of the CT sim is to eliminate the need to perform conventional treatment simulation by performing simulation on a computer generated or virtual image of the patient. In terms of functionality, the CT sim consists of a conventional CT scanner, a dedicated computer workstation, and some form of laser patient marking system that is used to define either a portal field shape or tumor volume isocenter on the patient's skin surface. Figure 1 is a diagram of a commercial CT sim developed by Picker International Inc. which employs either a conventional or spiral CT scanner, a dedicated computer workstation (virtual simulator) and an external laser patient marking system. Other systems produced either commercially[2,3,4] or as a research tool[5,6,7,8] use components similar to those described above but differ largely in the laser marking technologies employed. In particular, systems have been designed that contain laser marking systems mounted inside the gantry[2,4,5] or affixed to the outside of the scanner gantry.[6,7]

ACCEPTANCE TESTING

Acceptance testing is an important part of the purchase and installation of any new piece of equipment. It is the point at which the user has the opportunity to evaluate the system to ensure that it meets the initial specifications that the manufacturer and purchaser have agreed upon and probably the only time that the user may be able to ensure that the system is modified if the initial specifications are not met. As a result, acceptance testing should: (a) evaluate the performance of each component of the system individually, (b) compare the values obtained in (a) to ensure that they meet specifications, and (c) initiate actions to rectify errors detected in (b). A thorough protocol should also pay particular attention to the evaluation of non-mechanical parameters. This is especially true for the software used to perform the virtual simulation process. Although manufacturers of software products are required to adhere to practices that ensure that their product contains few if no errors, it has been shown that large computer programs that contain 30,000 or more lines of code can have over 100 errors, even after release from the manufacturer.[9] The need to perform a thorough evaluation of new equipment is also important to ensure that the equipment complies with both state and federal regulations.

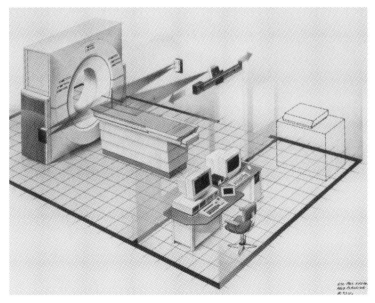

Figure 1. CT scanner and laser marking system on the Picker CT simulator system.

The amount of literature describing acceptance testing protocols for CT simulators is sparse.[10,11] For conventional CT scanners an excellent handbook exists to describe the steps needed to accept and commission a scanner.[12] If the computer hardware and software used to perform virtual simulation is considered part of a larger classification of systems to perform treatment planning, then some literature exists to describe acceptance testing protocols.[9,13] Because of the scarcity and heterogeneity of acceptance testing documents, the purpose of this section is to describe in detail the steps necessary to perform an acceptance testing and commissioning of a CT simulator. The protocol is broken down into three sections relating to the main components of the scanner. Section I outlines the protocols relating to the CT scanner, section II, the laser patient marking system, and section III, the virtual simulator computer and software.

SECTION I. ACCEPTANCE TESTING THE CT SCANNER

I.1 Test Phantoms

A number of test phantoms are required to perform acceptance testing of the CT scanner. As a standard, a water phantom similar to that described in AAPM report number 1[14] should be used. The phantom contains several test patterns inserted into a water filled cylinder. It is designed to evaluate image noise, spatial resolution, low and high contrast resolution, absorbed dose, field size effect, slice thickness, alignment, and mAs linearity. Typically, as part of the purchase of the CT scanner, the manufacturer will provide the same phantom or a modified version of it that contains some or even all of these test patterns. Other equipment necessary for performing acceptance testing is also well documented.[12]

I.2 What to Test

Unlike other components of the CT simulator, acceptance testing and commissioning procedures for a conventional CT scanner are well documented.[12] *Specification and Acceptance Testing of Computed Tomography Scanners*[12] is an in-depth review of the various aspects of commissioning of a CT scanner and lists extensively the various acceptance tests and the associated equipment required. This document also provides information for data such as electron densities of various materials and room shielding requirements. Table 1 is a summary of the tests necessary to perform acceptance testing of a conventional CT scanner as described herein. The table lists the type of test required, its priority, and comments explaining the impact of each test on the performance of the CT scanner.

The manufacturer of the CT scanner should also provide documentation describing any acceptance testing that was performed at the time of manufacture. This information should describe the characteristics of the scanner x-ray system and parameters that define the image quality of the reconstructed CT slice such as image noise, low and high contrast resolution, and spatial uniformity. The manufacturer will also provide software routines or allow the user to evaluate these image parameters at the time of installation. These tests can be integrated into the routine quality control program for the scanner.

I.3 Measurement Tolerances

Data generated from the test protocols listed in Table 1 should be compared to measured

Table 1. Tests Required to Perform Acceptance Testing of a Conventional CT Scanner*

CT Acceptance Test	Priority	Comments
Localization light accuracy	Essential	Affects clinical utility. Can be combined with table/gantry tilt test.
Alignment of table to gantry	Optional	Useful if misalignment is suspected
Table gantry tilt	Optional	Implications in QCT†
Slice localization from radiographic (scout) image	Alternative to table/gantry tilt	Requires special test object
Table increment	Essential	Implications in image quality and patient dose
Radiation profile	Essential	Affects patient dose
Sensitivity profile	Essential	Measures accuracy of image slice width, implications in image quality
mAs linearity	Optional	Some implications in noise scaling with dose
kVp accuracy	Optional	Error can be problematic in QCT† but is difficult to measure
Image noise	Essential	Can substitute with low contrast resolution test
Field uniformity	Important	Implications in image quality and QCT†
Scan field dependence (for QCT†)	Optional	Implications for QCT† with some scanner designs
High contrast resolution — resolution pattern	Essential	Requires resolution test pattern from CT vendor or third party
High contrast resolution — MTF	Alternative to resolution pattern	More quantitative than resolution pattern test but requires MTF tool and interpretation software
Low contrast resolution	Alternative to image noise	Requires special phantom, can replace noise measurements but requires observer performance test.
Display and hard copy image quality	Essential	Requires SMPTE pattern as CT image data file or external video generator
Radiation dose	Essential	Requires dosimetry phantoms and CT ion chamber or TLD system

*Table reproduced with permission, American Association of Physics in Medicine, 1994.
†QCT = Quantitative CT.

specifications defined by the AAPM Task Group 2 report.[12] In the absence of tolerances defined by this protocol, the data should be compared to the original bid specifications or manufacturer's specifications.

SECTION II. CT SCANNER-VIRTUAL SIMULATOR INTERFACE

Before virtual simulation can be performed, axial CT data must be transferred from the CT scanner to the virtual simulator. It is thus important to evaluate the transfer link between the two. Commercial CT simulators will typically have a dedicated link between the two systems but this may not be true for all systems. Some systems may require interfacing through an existing local area network (LAN) or may require a specifically tailored solution. Systems that have been developed in-house may have less efficient means of sending images from the scanner to the virtual simulator. As a result, the speed and accuracy of the transfer of images between the two systems should be evaluated under a variety of situations.

Initial Data Transfer

The ability to transfer CT data from the scanner to the virtual simulator should be evaluated by sending a CT study containing a few slices to the virtual simulator. This protocol should be incorporated daily, as part of the 'warm-up' protocol for the scanner.

Speed of Data Transfer

The speed of data transfer between the two systems should be determined for a variety of operating conditions. Evaluation should involve transferring a large CT study of typically greater than 60 CT slices between the two systems and recording the time taken to transfer the data. This should be repeated at least five times during a normal treatment day. For example: (1) at the time of morning warm-up, before patient simulation has begun, (2) two to three hours after the initial warm-up, (3) at the nominal lunch time period (typically between noon and 1:00 pm), (4) two to three hours after lunch, and (5) after normal treatment hours. Differences between the five periods of the day will indicate the effect of LAN traffic on the transfer rate of data if a nondedicated link is used. These results should be compared to acceptable benchmarks determined by the user.

Cyclical Redundancy Checks (CRCs)

CRCs are simply data integrity checks to ensure that data transferred from one point to another does not become corrupted during the transfer process. Manufacturers of the CT sim or LAN should be contacted to ensure that CRCs are part of the data transfer protocol.

II.1 Measurement Tolerances

Table 2 lists measurement tests and acceptable tolerances for ensuring data integrity in the image transfer process between the CT scanner and virtual simulator.

SECTION III. LASER PATIENT MARKING SYSTEM

The purpose of the laser patient marking system is to transfer the coordinates of the tumor isocenter, derived from the contouring of the CT data set, to the surface of the patient.

Table 2. Acceptance Testing Protocols for the CT Scanner Component of the Virtual Simulator

Acceptance Test	Acceptable Tolerances
Initial CT data transfer	Successful transfer of all CT data
Transfer of axial CT data to virtual simulator	Transfer speeds should be within tolerances for all times of the day as defined at time of commissioning
CRC evaluation	Data integrity verified by CRC check

Two types of laser marking systems are used to define the portal field aperture and isocenter on the patient's skin. The first system involves a gantry mounted laser that scans the portal field and defines the isocenter on the patient's skin.[5] The second system, as depicted in Figure 1, consists of a wall mounted, moveable sagittal laser and two stationary lateral laser systems. The two lateral lasers each consist of two separate lasers that define a vertical and horizontal line. The intersection of the two represent the lateral isocenter position. The acceptance testing protocol used to test the laser marking system described here will focus on the second of these two systems but is equally applicable, with the appropriate modifications, to other systems.

III.1 Test Phantoms

Figure 2 shows a commercially available test phantom (therapy alignment gauge, manufactured by TEL-ALIGN™) that has been modified to evaluate the wall mounted system shown in Figure 1. Two crosses have been inlaid into the top face of the phantom, 4.0 centimeters on either side of its central axis. The crosses were constructed from 1.02 mm diameter solid brass wire and are inlaid so that the center of the wire is level with the top face of the phantom. The center of each cross also falls along a line between the centers of the two side markers of the phantom.

III.2 What to Test

Acceptance testing of the laser patient marking system should involve an evaluation of the spatial linearity of the sagittal laser drive mechanism, the alignment of the sagittal and lateral lasers with the center of the CT scanner gantry, and the line width of each laser.

Sagittal Laser System

Three components of the sagittal laser system should be evaluated. These include: (1) coincidence of the sagittal laser zero position and gantry center, (2) the deviation of the laser from the vertical, and (3) the spatial linearity of the laser drive mechanism. The coincidence of the sagittal laser and radiation field center can be evaluated using the test phantom shown in Figure 2 and is described in the following section. The divergence of the laser system from the vertical can be evaluated using the phantom shown in Figure 2 and a spirit level. First, the

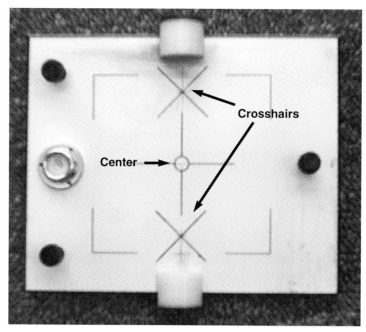
Figure 2. Laser alignment test tool.

test phantom should be placed on the CT table with the center of the phantom coincident with the sagittal laser at the zero position. The phantom should have a center mark for easy alignment. Next, place the spirit level on its end on the test phantom with the edge of the level at the center of the phantom. Level the phantom using the leveling screws. A nondiverging laser will track the edge of the spirit level. If the laser diverges from the vertical it should be clearly evident. The vertical divergence should ideally be no greater than 1°. The spatial linearity of the laser drive mechanism can be checked by taping a ruler to the table of the CT scanner parallel to the direction of motion of sagittal laser. Starting at the zero position, the laser should then be moved in 1.0 cm steps to both limits of the laser drive position and back to zero. Spatial nonuniformities should manifest themselves as deviations between the position of the respective centimeter marks on the ruler and the expected position of the laser. The difference between the measured distance as defined by the ruler and laser should ideally be less than 1.0 mm at the two maximum positions of the laser.

Calibration of Gantry, Lateral, and Sagittal Laser to Gantry Center

The alignment of a gantry mounted laser to the radiation field center as well as the alignment of the sagittal and lateral lasers to the gantry center can all be evaluated using the phantom shown in Figure 2. First, the phantom should be placed on the table and leveled. The CT table is then raised or lowered until the horizontal laser of each lateral laser set is coincident with the horizontal isocenter marker on each side of the phantom. If the two horizontal lasers are not coplanar then the table can be raised or lowered so that one of the two lasers matches the mark on the phantom. The couch is then moved into the gantry so that the gantry laser intersects the two crosses at their centers and the sagittal laser is coincident with the center mark of the phantom when it is at its zero position. The coincidence of the radiation field and gantry laser is then evaluated by taking an axial slice through the phantom using the smallest slice thickness allowable on the scanner (typically 2 mm). If the radiation field and laser are aligned each cross will appear as a single, small circle as the radiation fan beam is passing directly through the center of each. When misalignment of the laser and radiation beam occurs the beam will no longer cut through the center of one or both of the crosses. The reconstructed image at the location of the cross will show two circles rather than one and the separation of the circles represents the degree of misalignment. Figure 3(a) is an example of a situation when the radiation field and gantry laser are aligned while Figure 3(b) shows the situation when the gantry laser has drifted.

To determine if the sagittal laser and horizontal components of the lateral lasers are coincident with the center of the CT scanner gantry (i.e., radiation field), one of the measurement tools provided on the CT scanner software should be used to measure the coordinates of the centers of the two crosses of the test phantom shown in Figure 2. Because the center of each cross is equidistant from the center of the phantom, the absolute value of the horizontal coordinates of the two cross centers should be equal. In the case of the phantom shown in Figure 2, the horizontal coordinates of the two centers should be ± 4 cm. When this is true, the horizontal center of the phantom and gantry are coincident indicating that the zero position of the sagittal laser is coincident with the gantry center. The coincidence of the horizontal components of the lateral lasers and the gantry center is verified by ensuring that the vertical coordinates of the two cross centers are zero. If the gantry, sagittal, and horizontal components of the lateral lasers are aligned, the vertical component of the lateral lasers can

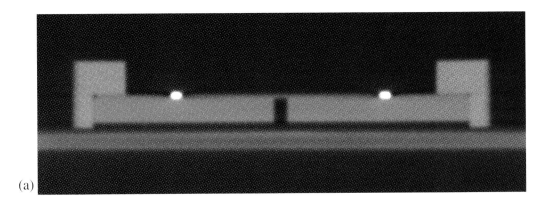

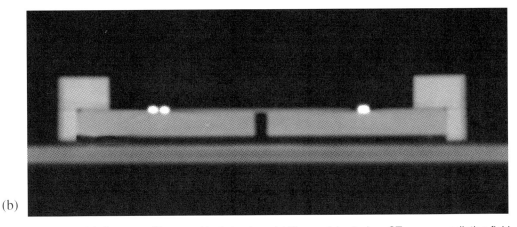

Figure 3. Transaxial alignment of laser and test phantom. (a) Transaxial cut when CT scanner radiation field and gantry laser are aligned. (b) Transaxial cut when CT scanner radiation field and gantry laser are misaligned. The diameter of the center of the cross on the right is approximately 3 mm.

be checked by moving the couch a predefined offset distance back out of the gantry. This offset distance is the distance from the center of the CT scan circle to the position of the lateral laser's isocenter and should be stored in the software of the virtual simulator. The offset value can be usually be obtained by contacting the service engineer of the CT simulator or by checking the user manuals of the system. The table should be moved this distance and misalignments will manifest as shifts of the laser relative to the vertical axis of the side isocenter markers of the phantom.

Line Width of Lasers

The line width of the lasers should be checked to ensure that they comply with manufacturers' stated tolerances.

III.3 Measurement Tolerances

Table 3 lists suggested recommended acceptance tests and tolerances for the laser marking system. It should be noted that these values were derived from quality assurance protocols developed at the Fox Chase Cancer Center. If individual systems are unable to meet these tolerances, then they should be revised accordingly.

SECTION IV. VIRTUAL SIMULATOR

The term "virtual simulator"[15] in this context refers to a group of software packages that allow the user to define and calculate a treatment isocenter and then simulate a treatment using digitally reconstructed radiographs (DRRs). Hence, the virtual simulator can be comprised of two main components: (a) the tumor or target localization package that allows the definition of a target volume to be performed and the transfer of these coordinates to the patient skin surface and (b) the virtual simulation package that generates a DRR which is then used to perform treatment simulation. The philosophy behind the development of this acceptance testing protocol is to evaluate the individual components of the virtual simulator to obtain a measure of the overall performance of the system. With this in mind the protocol described below details a series of tests to evaluate the essential components of the virtual simulation software.

Table 3. Acceptance Testing Protocol for Patient Laser Marking System

Component	Acceptance Test	Acceptable Tolerances
Sagittal laser system	Alignment of laser and center of radiation field	Error no greater than 1.0 mm
	Vertical divergence	Less than 1° over 100 cm
	Spatial linearity of motor drive	1.0 mm at maximum positions
Gantry laser/radiation field coincidence	Laser isocenter check	Less than 1.0 mm difference between two isocenters
Lateral laser vertical offset	Laser isocenter check	Less than 1.0 mm
Lateral laser longitudinal offset	Laser isocenter check	Less than 1.0 mm

IV.1 Test Phantom

A quantitative assessment of the virtual simulator cannot be performed without a CT data set of an object whose physical dimensions are accurately known. The simplest and most efficient method of meeting this criteria is to develop a phantom that can be used to generate the CT data set upon which acceptance testing of the virtual simulator can be performed. Figure 4 is an example of such a phantom developed at the Fox Chase Cancer Center. The phantom consists of a 15 cm^3 block of high density polystyrene with four separate test patterns inlaid into the top face of the phantom. Figure 4(a) is a photograph of the phantom while 4(b) is a line drawing of its top face showing the four patterns. The patterns enable modulation transfer function, contrast detail, spatial linearity of the DRR as measured by the registration reconstruction error, and ray line divergence accuracy of the DRR algorithm. The phantom itself provides the fifth test pattern as it allows contouring verification. Of these patterns, two are not necessary for acceptance testing but are instead used for quantitatively evaluating DRRs and has been described previously.[16] Table 4 lists the individual properties of the phantom and inlaid test patterns.

IV.2 What To Test

Target Localization

The definition of a target volume is addressed in the chapters on specific sites and treatment planning. Contouring a target volume involves manual or semiautomated packages in the virtual simulator. The isocenter of the target volume is then calculated and the position of the isocenter is transferred to the patient by the laser system.

Target Contour Verification

The acceptance testing procedure should involve the evaluation of the contouring capabilities of the virtual simulation program. A CT scan of the test phantom should be obtained and transferred to the virtual simulator. Contouring software should have an option to manually contour an object using a light pen or mouse as well as some form of semi automated routine based on the thresholding of CT numbers to determine the contour boundary. Both of these components should be evaluated by contouring the perimeter of the phantom on each CT slice. In an ideal case, the error should be no greater than the pixel size. An estimate of the contouring accuracy can be obtained from the difference in the physical area of the phantom through which the CT cut is taken and the area defined by the contour. The software should contain a measurement tool for determining area.

Isocenter Calculation

The accuracy of the isocenter calculation routine should be determined by contouring the phantom and calculating the isocenter of the target. The three coordinates of the isocenter can be evaluated by initially contouring only the central axis of the phantom. If the central axis slice has been calculated correctly, the isocenter as defined by the lateral and anterior lasers should coincide with the origin of the grid patterns on the anterior and lateral faces of the phantom. Placing more than one contour on the CT study of the phantom will allow the longitudinal accuracy (superior to inferior direction) to be evaluated.

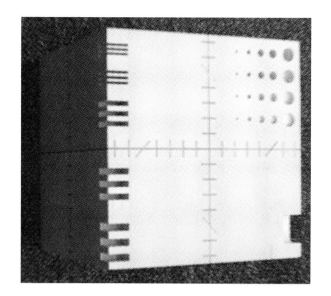

Figure 4. Virtual simulator test phantom. (a) Photograph of test phantom. (b) Line drawing of test patterns inlaid into the top face of the phantom. The patterns allow modulation transfer function (MTF), contrast-detail (CD), registration reconstruction error (RRE), and ray line divergence (RLD) to be evaluated.

Table 4. Test Patterns Found on the Phantom Shown in Figure 4*

Test Pattern	Construction	Purpose
Phantom block	15 cm^3 cube of high density polystyrene	Evaluate contouring component of virtual simulator
Central axis grids	1.016 mm brass wire	Evaluate spatial linearity of DRRs
Ray line divergence	Three 1.0 mm length, 1.016 mm diameter brass wires	Evaluate ray tracing accuracy of DRRs
Line pair patterns	1.5 cm rectangular slits of varying thickness	Measure DRR modulation transfer function
Contrast detail	Diameter and depth of holes varied	Measure DRR low contrast

*Test patterns 4 and 5 are not used for acceptance testing but are instead used to evaluate digitally reconstructed radiographs.

Isocenter Movement

If an isocenter is not defined using the contouring mode described previously, then often the position of the isocenter can be defined using a mouse. In this instance the operator indicates the isocenter at a CT slice and location within that slice. The virtual simulator will then designate this location to the skin surface using the laser systems. This component can be evaluated by moving the isocenter to various CT slices and landmarks within the slice. Ideally, a specific landmark within the slice should be chosen, for example, the various tick marks of the central axes of the grid pattern. The coincidence of the selected position of the isocenter and the actual position as defined by the lasers on the phantom can then be verified.

DRR Verification

The DRR is the digital equivalent of the conventional simulation film and is discussed in detail in Chapter 5. The purpose of the DRR is threefold. It must: (1) provide adequate target volume definition, (2) allow the visualization of contoured structures, and (3) and be able to visualize bony landmarks for treatment verification purposes. The DRR could also be used as a legal record of the treatment. Acceptance testing of DRR images should include evaluation of the various aspects of the reconstruction algorithm. In particular, the ray tracing accuracy and interpolation routines should be evaluated to ensure that the ray-line divergence is correct and that the DRR is free from spatial distortions.

Ray Line Divergence

The ray tracing component of the DRR reconstruction algorithm can be assessed using the ray line divergence pattern listed in Table 3. The pattern as shown in Figure 5 consists of three brass wires of 1.02 mm diameter and 2.0 cm length. The first of the three is located a distance of 5.0 cm from one of the central axes of the top face of the phantom with one end touching the phantom edge. The two other rods were placed on the opposing face of the phantom at distances of 5.38 and 6.25 cm from the central axis of the bottom and parallel to the rod on the top face. These rods are positioned so that the distance from each rod to the side of the bottom

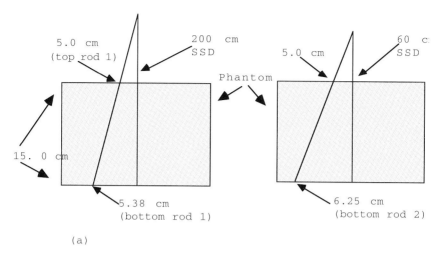

Figure 5. Ray line divergence tool. (a) Coincidence of top and bottom rods when an SSD of 200 cm is set. (b) Coincidence of top and bottom rods when an SSD of 60 cm is set.

face was 2.0 cm. The distances of 5.38 and 6.25 cm were chosen so that when the distance from the virtual radiation source to the top face of the phantom or the source to skin distance (SSD) is set to 60 cm then the rod on the top face and that at 6.25 cm on the bottom will fall along the same ray line if the ray tracing component of the algorithm is working correctly. Similarly, if an SSD of 200 cm is set then the rod on the top and the bottom at 5.38 cm will be coincident. Errors in the ray tracing algorithm will manifest themselves as shifts of the top rod with respect to the bottom as seen on the DRR. It is important to note that the SSD distances of 60 and 200 cm represent the minimum and maximum values for these parameters as determined by the manufacturer of the virtual simulator software. If other software manufacturers provide other limits, the rod distances should be modified accordingly. Figure 5(a) shows how a single ray line traced through the top rod will pass through the bottom at 5.38 cm when an SSD of 200 cm is set while (b) shows the situation for an SSD of 60 cm.

Spatial Linearity

The spatial linearity of a DRR can be easily evaluated using the spatial linearity grids that appear on the two orthogonal faces of the phantom. If the virtual simulation software contains measurement tools such as a ruler function, the distances between the tick marks located on the two faces can be measured. Spatial distortions of the DRR will manifest themselves as deviations from the true physical distances. If a measurement tool is not available, but a graphical overlay of the field size is, spatial distortions can be quantitated by measuring the differences between the grid pattern and field border for various field sizes.

Contour Verification Using Beam's Eye View Mode

Contours defined earlier as part of the target localization component should be evaluated to ensure that, like DRRs, the contours project correctly and are free from spatial distortions. Divergence of contours can be evaluated using a technique similar to that described in the section on ray line divergence and Figure 5. In this instance, a single axial contour of the phantom is displayed a fixed distance from the isocenter of the field. By setting the isocenter at the top face of the phantom and varying the SSD, the distance of the contour on the opposing face of the phantom from the central axis can be calculated using similar triangles. The difference between the calculated and measured distances is an estimate of the errors in the contour divergence. The spatial linearity of the contours can be evaluated by setting the isocenter to the top face of the phantom and selectively turning individual contours on and off. The coincidence of the contours and grid pattern will then reveal if any spatial distortions exist.

Verification of Virtual Simulation Variables on DRR and Contour Accuracy

As one purpose of the virtual simulator is to emulate conventional simulation, the virtual simulator must contain the same functionality as a conventional simulator. Table 5 is a summary of essential features of any virtual simulation program.

Virtual Field Size

Using the phantom described in Figure 4, field size accuracy can be assessed by comparing the coincidence of the tick marks of the linearity pattern with the field borders. Both rectangular and square sizes should be evaluated from the minimum to maximum allowable sizes. If asymmetric field sizes are allowable then these should also be evaluated.

Virtual Gantry Rotation

The test phantom shown in Figure 4 allows the evaluation of gantry rotations for orthogonal planes and can be easily evaluated by setting the isocenter of the DRR at the center of the phantom and rotating the virtual gantry in 90° steps. The coincidence of the field isocenter and origin of the grid pattern after a rotation will give an estimate of the rotational accuracy of the DRR algorithm. A range of gantry angles, including those that are not 0° or multiples of 90°, should be evaluated to eliminate the possibility of errors being undetected because of zero multiplication in the rotation matrix.[9] Also, small angle rotations about any of the major axes should be tested to ensure that small rotations do not produce erroneous results.

Virtual Collimator Rotation

Virtual collimator rotation should be evaluated for a range of angles similar to those described in the previous section. A particularly useful technique is to first choose a collimator rotation angle, then adjust the field size so that one of the central axes of the beam passes through two corners of the field such that it divides the field into two rectangles. The length of the hypotenuse of the rectangle can be calculated from the dimensions of the field. The cal-

Table 5. Essential Components of a Virtual Simulation Program

Component
Virtual Gantry Rotation
Virtual Collimator Rotation
Virtual Table Rotation
Virtual Field Size
Virtual Isocenter
Magnification

culated length can be compared to the actual length. The actual length is given by the intersection of the field corners and the central axis of the beam bisecting the field.

Virtual Table Rotation

Special attention should be paid to the ability of the virtual simulation system to determine if a physically impossible treatment set up is detected. For example, if a couch and gantry angle of 90° will result in a collision, the user should be notified of such an error. The angles of rotation should be made in small increments (15° or less). The accuracy of virtual couch rotations should be assessed using the same technique described for virtual collimator rotations.

Isocenter

The ability to shift the isocenter from one known position to another on a DRR should be determined. Using the phantom described in Figure 4 the position of the isocenter should be shifted to various landmarks on the phantom. For example, the rotated tick marks at distances of 5 cm along each of the grid patterns can serve as positions to shift the isocenter to and determine the spatial accuracy of this component of the algorithm.

Film Magnification (Zoom Function)

A range of physical magnifications typically used for conventional simulation images should be evaluated. This should involve generating film hard copies at various magnifications and verifying the actual magnification on the films.

Combinations of Gantry, Collimator, Table Rotations, Field Size, and Isocenter Shifts

Various combinations of the six parameters described above should be evaluated to determine if any errors exist as a result of the combinations.

IV.3 Measurement Tolerances

Table 6 summarizes tolerance levels for the evaluation of the virtual simulation component of the CT simulator. This table lists the various components to be evaluated, the appropriate test, and the acceptable tolerance for each. It should be noted that many of the tolerances stated in Table 6 originate from limits that were deemed acceptable at our institution as well as the physical limits of the data sets used (e.g., pixel size, slice thickness, etc.). Individual institutions should accept or revise these values based on their equipment and its proposed use.

V. TEST CASES

The accuracy of the virtual simulator depends upon the resolution of the data set. As conventional scanners allow the slice thickness, separation between slices, and reconstruction area to be varied, the virtual simulation program should be evaluated for various combinations of these scan parameters. As the number of possible combinations can be quite large, the approach at the Fox Chase Cancer Center has been to evaluate combinations of these variables that are used on a routine clinical basis. Table 7 is a summary of the combinations of these parameters that were evaluated for the CT simulator installed in our department.

Table 6. Acceptance Testing Protocol for Virtual Simulator

Virtual Simulator Component	Acceptance Test	Tolerance
Target localization	Target contour verification	Error no greater than 1 pixel
	Isocenter calculation	No greater than half slice thickness along scan axis and one pixel along the transaxial axis
	Isocenter movement	Same as isocenter calculation
Digitally reconstructed radiograph (DRR)	Ray tracing	Angular divergence less than $1°$ over 100 cm
	Spatial linearity	Error no greater than 1.0 mm over 10 cm distance
Contour verification	Contour divergence	Error no greater than $1°$ for all field sizes at 100 cm SSD
	Spatial linearity	Error less than 1.0 mm between grid spacings over 10.0 cm
Virtual simulation Parameters	Gantry rotation (G)	Less than $1°$ over $90°$ interval
	Collimator rotation (C)	Less than $1°$ over $90°$ interval
	Table rotation (T)	Less than $1°$ over $90°$ interval
	Field size (F)	No greater than half slice thickness along scan axis and pixel size along the transaxial axis
	Isocenter (I)	No greater than half slice thickness along scan axis and pixel size along the transaxial axis
	Magnification (M)	Less than 1% over the range of magnifications of 0.75 to 1.5
	Combinations of C, T, F, I, & M	Combined tolerances of individual tests

Table 7. Values of Reconstruction Circle, Slice Thickness, and Separation That Were Used to Generate CT Data Sets in Order to Perform Acceptance Testing of the CT Simulator

Reconstruction Circle	Slice Thickness (mm)	Slice Separation (mm)
Full (48 cm diameter)	5	5
Full 5	3	5
Full 2	2	5
Half (24 cm diameter)	5	5
Half 5	3	5
Half 2	2	5

Table 8. Quality Control Schedule for Various Components of the CT Simulator

CT Simulator Component	Test	Frequency*
CT scanner	CT noise uniformity	I, W, M, Y
	CT number uniformity	I, W, M, Y
	CT low contrast resolution	I, M, Y
	CT MTF measurement	I, M, Y
	Display and hard copy uniformity	I, M, Y
	Table gantry tilt	I, Y
	Scout localization accuracy	I, Y
	Slice thickness and sensitivity	I, Y
	Spatial reconstruction uniformity	I, Y
	HVL measurements	I, Y
	mAs linearity	I, Y
	Dose measurements	I, Y
	Transfer of CT data to virtual simulator	D
	Speed of CT data transfer	I, M
Laser system	Alignment of radiation center and gantry laser	I, D
	Alignment of sagittal laser	I, D
	Alignment of lateral laser	I, D
	Alignment of laser and center of radiation field	I, D, W, Y
	Vertical divergence	I, D, W, Y
	Spatial linearity of sagittal laser	I, M, Y
Virtual simulator	Computer hardware	I, after hardware changes
	Computer software	I, after any software change/upgrade

*I indicates initially; D, daily; W, weekly; M, monthly; and Y, yearly.

VI. GENERATING A QUALITY CONTROL PROTOCOL

After the acceptance testing of the CT simulator, a quality control protocol should be developed to check critical tolerances of various components of the system. In particular, components such as the CT scanner and laser marking system should be checked routinely while the hardware and software components of the virtual simulator should be checked at the time of acceptance testing and every time each component is upgraded (or anytime a new function requires reinstallation of the software). Table 8 is a description of the various components of the CT simulator that should be checked and their frequency. Data relating to the frequency of testing of a CT scanner was taken from guidelines published from the New York Department of Health, Bureau of Environmental Radiation Protection.[17] This data schedule of tests are performed in our department to ensure that the various components of the CT simulator comply with their allowable tolerances.

REFERENCES

1. Webb S. Non-standard CT scanners: their role in radiotherapy. Int J Rad Oncol Biol Phys 1990;19:1589-1607.

2. Smith R, MmS LJ, Steidley KD, Kohut HT. Clinical patterns of use of a CT-based simulator. Med Dosim 1987;12:17-22.

3. Sims C, Root G, Skinner A, Galvin J. The use of digital reconstructed radiographs in CT simulation. Med Phys 1993;20:1298.

4. Heidtman CM. Clinical applications of a CT simulator: precision treatment planning and portal marking in breast cancer. Int J Rad Oncol Biol Phys 1990;15:113-117.

5. Ragan DP, He T, Mesina CF, Ratanatharathorn V. CT-based simulation with laser patient marking. Med Phys 1993;20:379-380.

6. Nishidai T Nagato Y, Takahashi M, Abe M, Yamaoka N, Ishihara H, Kubo Y, Ohta H, Kazusa C. CT simulator: a new 3-D planning and simulating system for radiotherapy: part I. description of system. Int J Rad Oncol Biol Phys 1989;18:499-504.

7. Nagata Y, Nishidai T, Abe M, Takahashi M, Okajima K, Yamaoka N, Ishihara H, Kubo Y, Ohta H, Kazusa C. CT simulator: a new 3-D planning and simulating system for radiotherapy: part 2. clinical application. Int J Rad Oncol Biol Phys 1990;18:505-513.

8. Nagata Y, Nishidai T, Abe M, Takahashi M, Yukawa, Nohara H, Yamaoka N, Saida T, Ishihara H, Kubo Y, Ohta H, Inamura K. CT simulator: A new treatment planning system for radiotherapy. Int J Rad Oncol Biol Phys 1987;13:176.

9. Jacky J, White CP. Testing a 3-D radiation therapy planning program. Int J Rad Oncol Biol Phys 1990;18:253-261.

10. Pennington EC, Jani SK. Quality assurance aspects of a CT simulator. In Jani SK, CT Simulation for Radiotherapy. Madison, Wisconsin: Medical Physics Publishing, 1993, pp 147-160.

11. Pandya JV, Alappattu JC, Philip PC, Moore CW, Khan FR. Acceptance testing and quality assurance of a CT simulation system. Med Phys 1994;21:902.

12. Lin PJP, Beck TJ, Borras C, Cohen G, Jucius RA, Kriz RJ, Nickoloff EL, Rothenberg LN,

Strauss KJ, Villafana T. Specification and Acceptance Testing of Computed Tomography Scanners. AAPM Report #39. College Park, Maryland: American Association of Physicists in Medicine, 1983.

13. Van Dyk J, Barnett RB, Cygler JE, Shragge PC. Commissioning and quality assurance of treatment planning computers. Int J Rad Oncol Biol Phys 1993;26:261-273.

14. Judy PF, Balter S, Bassano DA, McCullough EC, Payne JT, Rothenberg L. Phantoms for performance evaluation and quality assurance of CT scanners. AAPM Report #1. College Park, Maryland: American Association of Physicists in Medicine, 1977.

15. Sherouse GW, Novins K, Chaney E. Computation of digitally reconstructed radiography for use in radiotherapy treatment design. Int J Rad Oncol Biol Phys 1990;18:651-658.

16. McGee KP, Das IJ, Schultheiss TE, Sims C. A quality assurance phantom for digitally reconstructed radiographs (DRRs). Med Phys 1994;21:902.

17. Guide for Radiation/Quality Assurance Program Computed Tomography Equipment. New York State Department of Health Bureau of Environmental Radiation Protection, 1992.

— Chapter 3 —
LOCALIZATION & FIELD DESIGN USING A CT SIMULATOR

Margie Hunt, MS

INTRODUCTION

Localization and field design are two of the most crucial steps in the treatment planning process. Undetected systematic errors introduced at these junctures can be propagated throughout the entire course of treatment. Successful treatment therefore depends in part on instituting procedures aimed at assuring accuracy and precision during these steps. The major goals of the localization and field design steps include:
- Acquisition of a patient data set
- Target and normal structure localization
- Definition and marking of a patient coordinate system often referred to as the triangulation points
- Design of the treatment field including the aperture
- Transfer of information to the treatment planning system
- Production of an image for treatment verification

During conventional simulation, x-ray fluoroscopy, radiographic films, field defining lights, laser alignment systems and various methods for obtaining patient anatomical information and entering it into the treatment planning system are all used to achieve these goals. With CT simulation, each goal can also be achieved, although the process is different. The patient data set is obtained and the localization is performed using CT images; fluoroscopy and film are replaced by DRRs; a laser alignment system is used for marking; and a virtual simulator software package is used for field design and the production of verification images. Transfer of all the necessary information to the treatment planning system is done through a direct computer link. A CT simulator can be thought of as performing two distinct functions: *physical simulation* which includes the first three goals and *virtual simulation* which encompasses the last three goals. One of the major achievements of CT simulation is the performance of all of these goals without recourse to a conventional simulator. The purpose of this chapter is to discuss in detail the steps involved in localization and field design using a CT simulator.

PHYSICAL SIMULATION

Acquisition of Patient Data Set

Obtaining an accurate, self-consistent patient data set is particularly crucial to CT simulation since the same data are often used for multiple localization and field design sessions. CT simulation procedures which document technical and imaging guidelines for each treatment site should be developed. As a minimum, the following issues should be addressed.

CT Scan Limits

During conventional simulation, physicians specify the treatment field borders relative to anatomical landmarks to guide the radiation therapist in the preliminary set up of the fields. In an analogous fashion, cephalad to caudad scanning borders should be specified prior to scanning the patient for CT simulation. The following issues should be considered when determining the scanning borders.

a. Treatment fields to be localized and designed — The scan region should encompass the treatment volume for all fields. Often physicians bring a patient to CT simulation primarily to localize the disease which will be treated during the cone down or boost. However, technical and cost efficiency will be maximized if the CT data set is used for localization and design of all the treatment fields.

b. Treatment planning requirements — Even if the CT data will be used solely for the design of a cone down or boost treatment, additional reasons for scanning well beyond the intended borders of these fields exist. Anatomical data extending beyond the limits of the cone down treatment volume are often needed for the production of composite isodose distributions, the accurate calculation of dose volume histograms or the calculation of isodose distributions for non-coplanar beam arrangements.

c. Quality of DRRs — In conventional simulation, the blades of the simulator collimator are opened beyond the treatment field borders to image surrounding anatomy. In an analogous fashion, patients should be scanned beyond the actual treatment volume to image surrounding anatomy on the digitally reconstructed radiographs (DRRs).

Imaging Parameters

General imaging parameters should be defined for each anatomical region. Imaging protocols supplied by the manufacturer are optimized for diagnostic rather than CT simulation applications. Scan slice thickness and spacing must be evaluated on a site specific basis while considering volume averaging effects, target volume geometry, overall scan and image reconstruction time, and DRR quality. Localization of target volumes and normal structures which are rapidly changing in the cephalad-caudad direction require smaller slice thicknesses and spacing. The improvement in DRR quality as a function of slice thickness and spacing must be weighed against increases in total scan and image reconstruction time and image storage requirements. Patient movement as a result of long scan times leads to inaccurate localization which often becomes evident only upon viewing the DRRs. The introduction of spiral CT scanners on CT simulators will alleviate some of these problems.

Use of Contrast

Image enhancement through the administration of contrast material should be evaluated for each treatment site. Various contrast agents can be used to differentiate tumor from surrounding normal tissues, delineate nodal volumes, and localize normal structures such as small bowel. The type of contrast and method of administration are similar to diagnostic applications although additional timing or positional constraints may exist.

One difficulty in the use of contrast for CT simulation lies in the frequent inability to visualize the material both on the CT images and the DRRs. The concentration of contrast material necessary to yield optimal CT images is often not sufficient for visualization on the DRRs. For example, oral contrast administered for visualization of the small bowel on CT images does not usually produce sufficient opacity for visualization on the DRR where it is necessary for field design. Contouring the contrast-containing structures prior to virtual simulation is one solution since the contours can then be overlaid onto the DRRs for field design. Another potential solution being developed by the manufacturer of the CT in our institution is to allow "weighting" of specific ranges of CT values during the calculation of the DRR. This allows selective enhancement or de-enhancement of structures.

Use of Surface Markers

Skin nodules, scars, grossly involved lymph nodes and previous radiation field borders are often delineated during routine simulation by radio opaque wires or beads. During CT simulation these areas can be contoured on the CT images and projected onto the DRRs during virtual simulation. The other option, which minimizes contouring time, is to delineate the area with small solder BBs or thin solder wire to allow visualization on the DRRs without producing unacceptable artifacts on the CT images.

Table 1 summarizes the scanning procedures for each major anatomic region as developed at Fox Chase Cancer Center. Such region specific protocols serve as the basis for scanning all patients. This information is augmented by discussion with the physician prior to bringing the patient into the CT simulator. This table has been devised for our particular CT simulation system and is presented as an example. It should not be transferred to other systems without verification.

Two other issues related to obtaining the patient data set are worth mentioning. Typically, patients receive at least 70 axial scans during which they must remain in the treatment position. Prior to scanning, a preliminary set of alignment marks should be placed on the patient using the laser alignment system and the CT couch position should be recorded. Throughout the course of the CT simulation, the therapist may verify that the patient's position has not changed by returning the CT couch to the prerecorded position and verifying the alignment marks. This procedure is analogous to the preliminary patient marking done during conventional simulation.

Prior to scanning for CT simulation, a preliminary pilot view and a few preliminary axial scans of the patient may be necessary to ensure that the patient is positioned correctly and that the complete patient external contour will be within the scan reconstruction circle along the entire scan length. Complete external contours are necessary for accurate dose calculations.

Localization

Localization or contouring of the tumor, target, and normal structures can be a time consuming and tedious process which is greatly aided by an efficient, user friendly localization package. Desirable software features, not all of which are currently available on commercial systems, are outlined in Table 2. Figure 1 shows the image layout and main menu of the localization package on the Picker ACQSIM. Image manipulation and contouring features are accessed through the menu buttons on the left and right side of the screen, respectively. CT simulation systems are rapidly evolving and developments in image fusion (incorporation of MR, PET, SPECT, etc.), autocontouring, intelligent contouring, and three dimensional contouring methods will all improve CT localization in the future.

During localization, the physician defines and outlines the gross tumor volume (GTV),

Table 1. Fox Chase Cancer Center CT Simulation Scan Procedures

Site	Head, Neck, Brain	Thorax	Breast	Pancreas Gastric/Abd.	GU	GYN (Prostate)	Rectum
Immobilization	Aquaplast[1] or bite block	Alpha Cradle[2]	Alpha Cradle Angle board: (optional for CT)	Alpha Cradle	Alpha Cradle	Alpha Cradle	Alpha Cradle
Patient Position	H & N: Supine Brain: Discuss with MD prior to scan	Lung: Supine Esophagus: Prone	Supine	Supine	Supine	Supine	Prone
Scan Protocol	Trauma Brain	Thorax/Lung	Lung	Pelvis	Pelvis	Pelvis	Pelvis
Slice Thickness	2 mm	5 mm	5 mm*	5 mm	5 mm	5 mm	5 mm
Table Increment	2 mm	3 mm	3 mm*	3 mm	3 mm	3 mm	3 mm
Contrast	Brain: 100 cc Isovue 300[3] (at request of MD)	Lung: None Esophagus: 2T. EZ Paste[4] with 1/2 c. Readi-CAT[5]	None	1 Bottle Readi-CAT (1 hr prior), Additional 1/4 btl. before scanning	Urethrogram: 15 cc Reno-M- 30[6] IV Contrast: 25 cc Isovue 200	1 Bottle Readi-CAT (1 hr prior to scan), Vaginal marker: soaked in Readi-CAT	I bottle Readi-CAT (3 hr prior to scan)

Table 1. Fox Chase Cancer Center CT Simulation Scan Procedures (continued)

Site	Head, Neck, Brain	Thorax	Breast	Pancreas Gastric/Abd.	GU	GYN (Prostate)	Rectum
Scan Limits/ Special Notes	Brain: vertex to C2 vertebral body. Head and Neck: (except Larynx): 2 cm superior to sella to bottom of thyroid cartilage (or clavicle for SCV field). Larynx: External auditory meatus to bottom of clavicle if SCV field +	Lung: thyroid notch to diaphragm. Cervical Esophagus: Mastoid tip to carina. Thoracic Esophagus: Superior thyroid notch to carina. Lower esophagus: Superior thyroid notch to T12	Full chest: 5 mm slices, 3 mm increments above and below scar. Primary site and scar: 2 mm slice thickness, 2 mm increment	Diaphragm to top of iliac	Top of iliac crests to 2 cm below ischiums	Top of iliac crests to 2 cm below ischiums (check with MD prior to scanning)	Top of iliac crest to 2 cm below ischium

[1] WFR Aquaplast Corp, Wyckoff, NJ 07481
[2] Smithers Medical Products, Hudson, OH
[3] Squibb Diagnostics, Princeton, NJ 08543
[4] E-Z-EM, Inc. Westbury, NY 11590
[5] E-Z-EM, Inc. Westbury, NY 11590
[6] Squibb Diagnostics, Princeton, NJ 08543
[7] Beekley Corp., Bristol, CT 06010

Table 2. Virtual Simulation Software: Localization Features

Software Capability	Features
Viewing	Standard diagnostic CT image viewing capabilities
	Axial, sagittal, coronal, oblique CT images
	Pan, zoom capabilities for all images
	Posting for all structures
	Pilot (scout) images viewed concurrently with slices used for contouring.
	Position of contouring slices marked on pilot.
	Structures transposed on pilot using wireframe, outline or color-wash images
	Image fusion (CT, MRI, PET, SPECT image registration)
Structure/Contour Entry	Multiple entry methods (keyboard, mouse/trackball, light pen, etc.)
	Manual entry (rubberband, continuous data entry)
	Structure interpolation between slices
	Conventional autocontouring (single slice and slice range)
	Intelligent autocontouring using pattern recognition
	3D contouring
Structure/Contour Editing	Erase, stretch, copy, deletion, dilation, translation, rotation
	Editing features should apply to individual slice contours or to whole structures
	Editing of CT numbers
Structure Labeling	User-defined list of common structure name/color combinations
	Optional grouping of structures for the determination of isocenters or for the creation of new structures
Isocenter (Reference Point) Determination	Capability for isocenter determination based on single structure or structure groups
	Capability for isocenter determination based on single slice or whole structure
	Manual placement of isocenter (with or without structure definition)
	Capability for multiple isocenters
	Unambiguous transformation between the CT image and CT scanner coordinate systems
Statistical Features	Measurement (ruler)
	CT Number/VOI
	CT Histogram
Structure and Image Portability	Capability for transferring images, structures, isocenter coordinates, etc. to other departmental computer systems such as the treatment planning system

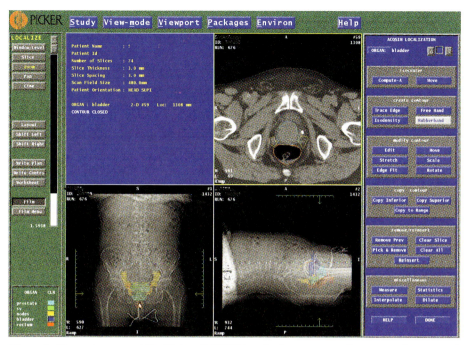

Figure 1. The main menu of the Picker ACQSIM localization package. Transverse images and pilot views are used for contouring. Software buttons for image manipulation are on the left. Contouring features are on the right.

clinical target volume (CTV) and planning target volume (PTV) as described in ICRU 50[1] and outlined in Chapter 7. All dose limiting structures must also be outlined by either the physician or planner.

Tumor and target definition is a complex issue which must be addressed at several levels. For the physician, two tasks exist: determining the areas to treat and actually delineating them on the patient's scans. For each patient, the physician must consider a myriad of information including site, stage, pathology, and performance status when deciding what anatomical regions or CTVs to treat. Once the CTVs have been defined, the physician must accurately contour them on the CT images often with guidance from surgeons and diagnostic radiologists.

The physicist's role in target definition is to guide the physician in defining the PTVs. The margin between the CTV and PTV must allow for the proximity of critical structures, organ motion, set up uncertainties as a function of immobilization, patient position and medical condition, and additional treatment uncertainties. At our institution, physicians routinely outline CTVs and specify margins to the PTV. The dosimetrist then produces the PTV from the CTV and has the physician review the final PTV making modifications as necessary. Many site specific measurements of set up uncertainty and organ motion have been published in the past few years, although methods for using these data to set margins and evaluate treatment plans are just emerging (see reference list in Appendix I). Until these methods are refined, physicians have no choice but to continue defining these margins semi-empirically, combining experience and training with these new data. Additional discussion of the issues involved in outlining tumor and target volumes can be found in Chapter 7.

Several CT simulation features can aid the physician and physicist in the difficult processes of defining GTVs, CTVs, and PTVs. Automatic dilation of structures is a convenient localization tool for creating PTVs based on the corresponding CTV and a uniform margin. Viewing anterior and lateral DRRs or pilot images with overlaid projections of the structures during the actual outlining is another helpful tool since it gives the physician immediate feedback on their work in a familiar context.

Patient Marking

Once the localization of the target volumes is complete, a coordinate system can be defined and marked on the patient using the laser alignment system. Typically the origin or "isocenter" of this coordinate system is specified at the center of the physician-defined CTV or PTV. The coordinates of the isocenter are calculated by the localization software and are specified in both the patient and the CT scanner coordinate systems. The CT couch is initially moved to the patient reference position to verify that the patient position has not changed over the course of the CT procedure. The couch and moveable lasers are then moved to the position of the origin and the transverse and sagittal laser projections onto the patient's surface are marked and tattooed. These three points can then be used as the triangulation points for treatment set up.

A good localization software package will offer the user flexibility in defining the patient coordinate system. It should be possible to arbitrarily place the origin (isocenter) at any point in the patient or to define it as the center of any structure or group of structures. The capability should exist for multiple isocenters and the coordinates of each one relative to the CT scanner coordinate system should be unambiguous and directly available to the user.

Once the patient coordinate system has been marked, the patient is free to leave the CT simulator since field design, the final step, can be completed in their absence.

VIRTUAL SIMULATION

Field Design

As discussed in Chapter 7, fields are commonly designed at one of three levels of complexity; all of which can be accomplished using a CT simulator. For any given patient, fields at any or all of the complexity levels may be used for treatment.

At the simplest level, physicians design fields based on radiographic anatomy alone without any CT localization of structures (e.g., whole brain fields). A DRR is created and overlaid with an electronic image of a field aperture and grid to produce an image analogous to that on a conventional simulator (Figure 2). The physician or planner can then modify any beam parameters including field position, angles, or size and shape of the aperture to create the desired field. Virtual simulation at this level of complexity is similar to conventional simulation.

At the intermediate complexity level, fields are designed using a combination of standard radiographic anatomy and projections of CT localized target volumes and normal structures on the DRRs (Figure 3). Standard beam portals such as opposed anterioposterior or lateral pairs which treat large nodal volumes in addition to the patient's gross disease are examples of fields which might be designed in this manner. Since non-involved lymph nodes are often difficult to visualize on CT, their position is often inferred from other structures. The physician may opt to define the field aperture in the nodal regions using the DRR radiographic anatomy and use the projected CT localized volumes to design a conformal aperture in the region of the gross disease.

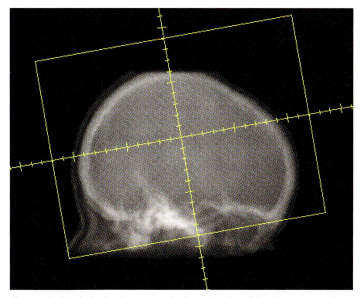

Figure 2. A whole brain portal designed using CT virtual simulation. Simple portals can be designed using the radiographic information from the DRR in a method exactly analogous to the conventional simulator.

At the highest level of field design complexity, all target volumes and dose limiting normal structures have been CT localized and are projected onto the DRRs (Figure 4). Complex fields, often at oblique angles or non-coplanar arrangements, are designed through a combination of geometric optimization using the virtual simulator and dosimetric optimization using the treatment planning system. The advantages of CT virtual simulation are most evident at

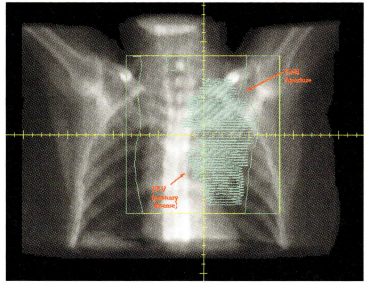

Figure 3. An anterior left lung field designed using a combination of radiographic anatomy visualized on the DRR and CT-localized structures projected onto the DRR. The radiographic anatomy was used to design the field aperture everywhere except in the vicinity of the PTV for the primary disease.

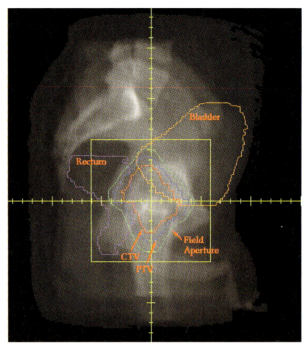

Figure 4. A right lateral prostate field designed using the projection of the CT-localized PTV and normal structures. Bladder, rectum, and the prostate CTV and PTV are all projected onto the DRR. The field aperture was designed to encompass the PTV with a margin of approximately 0.5 cm.

this level of complexity. Normal tissues not usually visible on a plain radiograph can be CT localized and projected onto the DRRs to aid in geometric optimization. Since DRRs for any field can be generated at any time, field modifications can be made throughout the planning process without having the patient return for more films. DRRs of any treatment field can be generated including those which could not normally be simulated (e.g., vertex fields).

Field design using a virtual simulator must function in essentially the same manner as a conventional simulator. Table 3 outlines the desirable software features for the field design package. Figure 5 shows the main menu of the Picker ACQSIM virtual simulation package. The DRR image with overlaid contour projections is shown as are menu buttons for all machine motions. In order to function as the sole localization and field design system within a department, some of the features not available today must be added. In particular, there must be a method to geometrically align or abut treatment fields either on the patient's surface or at depth. The standard beam arrangement used in the treatment of the intact breast and surrounding regional nodes for breast cancer offers a fine example of a complex geometry requiring field alignment. The inferior border of the supraclavicular field is often geometrically aligned with the superior border of the tangential fields and the posterior borders of the two tangential fields are made coplanar. Several radiographic methods such as those developed by Siddon et al.[2] exist for achieving the proper field alignments during conventional simulation. Capabilities such as room's eye view[3] must be added to the virtual simulation package to allow simulation of these types of beam arrangements.

Annotation or documentation on DRRs is another area where improvement in currently existing field design software packages is needed. Conventional simulator radiographs are

Table 3. Virtual Simulation Software: Field Design Package

Feature	Capabilities
Annotation	Institution and patient demographic information
	Field information including field name, all machine parameters, SSD, SAD
	Anatomical orientations
	Structure labeling
	Indication and labeling of ancillary devices such as wedges, compensators, MLC
	Machine icon showing position of machine head and treatment couch
Field Positioning and Design	Unambiguous coordinate system definitions
	Display of relative (wrt patient coordinate system) and absolute couch positions
	Capability to easily simulate fixed SSD fields
	Capability to easily determine the SSD for isocentric fields
	Capability to copy and oppose fields including apertures
	Collision detection capabilities
	Multiple beam parameter input methods (keyboard, mouse/trackball)
	Field parameter compatibility between CT simulator and treatment machines
	Room's eye view or similar method to view geometric relationship between multiple fields
	Capability to view the intersection of a beam with the patient's skin surface
Field Aperture Design	Automatic and manual aperture design
	Capability to define multiple apertures with different transmission values
	Multileaf collimator (MLC) and asymmetric jaw capabilities
Image Manipulation and Display	Contrast/brightness adjustments, etc.
	Image magnification to user defined values
	Structure posting
	Multiple methods for structure visualization (wire frames, outlines, color wash)
	Sufficient field of view to accommodate large treatment fields such as mantles
Field and Image Portability	Capability for transferring field parameters and images to other departmental computer systems such as treatment planning systems, block cutting devices, data information systems (record and verify, image display)

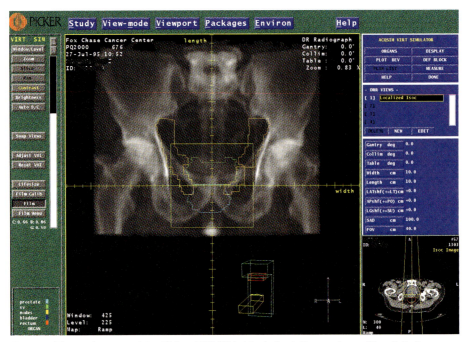

Figure 5. The main menu of the Picker ACQSIM virtual simulation package. The digitally reconstructed radiography is shown with contours of outlined structures and the field aperture overlaid. Machine motion buttons are shown on the right.

often labeled with anatomical orientations particularly for oblique or complex non-coplanar fields. The type, position, and orientation of various ancillary devices within the beam such as wedges and custom compensators is also often written on simulator films for quality assurance reasons. Similar capabilities must be developed for DRRs.

Information Transfer to the Treatment Planning System

After the fields have been initially designed, all pertinent information from the CT simulator can be transferred to the treatment planning system for dose calculations, optimization, and plan evaluation. This is most easily accomplished through a direct link to the planning system via a local area network (see Chapter 6). The information which is sent to the planning system at our institution at this point in time includes CT scans, structure contours, and isocenter positions. Only the CT images necessary for dose planning, usually a small subset of the entire data set, need to be imported to the planning system. In the near future we intend to have a bidirectional transfer between the planning system and the CT simulator which will allow the transfer of field parameters and apertures.

Production of Verification Images

At the completion of the dose distribution optimization, DRRs for the final treatment fields can be generated using the virtual simulator. Field apertures can be drawn on these images for use in the construction of custom blocks. The DRRs serve as a replacement for conventional simulation films in the comparison with portal images on the treatment machine.

Just as a technique chart is necessary to consistently obtain quality radiographs on a simulator, the window, level, contrast and brightness settings for optimum DRRs must be determined on a site specific basis. In general, we have found DRR quality for pelvic, brain, and head and neck sites to be sufficient to replace conventional simulation films. DRRs for abdominal sites such as pancreas are usually sufficient although for some cases conventional radiographs are still obtained because of insufficient spatial resolution in the area of the vertebral bodies. Patients being treated to thorax sites routinely go to conventional simulation because of insufficient resolution along the vertebral column and because of the inability to visualize both soft tissue and bony anatomy in a single DRR image. It is possible to produce multiple DRRs of the same field using different imaging parameters to enhance different types of tissues, however, this is quite time consuming. Therefore, our current policy is to require a conventional simulation of the final treatment fields for these patients.

SUMMARY

Figure 6 summarizes the major steps in the Fox Chase Cancer Center CT simulation process that were discussed in this chapter. Localization and field design using CT simulation is a logical progression in CT assisted treatment planning. Although the role of CT simulation is still evolving, several advantages over conventional simulation and CT assisted planning are becoming evident. Radiation oncologists can project the outlines of tumors and normal tissues contoured on the CT images directly onto DRRs and use this information for real-time geometric optimization and field design. Although this may be achievable with other virtual simulation packages, CT simulation offers the additional advantage of being able to define and mark the patient coordinate system during the CT session, thereby eliminating the need for

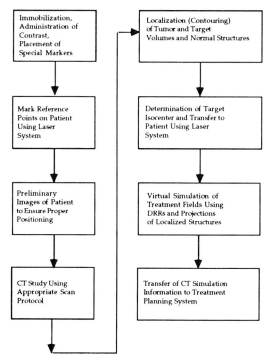

Figure 6. CT simulation process at Fox Chase Cancer Center.

patient visits to both the CT scanner and the conventional simulator. This can potentially improve cost efficiency and planning accuracy, since registration of the CT and simulator data sets is not necessary. Last but certainly not least is the advantage to the patient who, in a single visit to the CT simulator, provides the data set which will be used to design multiple "optimal plans" over the course of his or her entire treatment.

REFERENCES

1. ICRU Report 50. Prescribing, Recording, and Reporting Photon Beam Therapy. Bethesda, Maryland: International Commission on Radiation Units and Measurements, 1993.

2. Siddon RL, Buck BA, Harris JR, Svensson GK. Three-field technique for breast irradiation using tangential field corner blocks. Int J Radiat Oncol Biol Phys 1983;9:583-588.

3. Purdy JA, Harms WB, Matthews JW, Drzymala R, Emami B, Simpson JR, Manolis J, Rosenberger FU. Advances in 3-dimensional radiation treatment planning systems: room-view display with real time interactivity. Int J Radiat Oncol Biol Phys 1993;27:933-944.

4. Campostrini F, Garusi G, Donati E. A practical technique for conformational simulation in radiation therapy of pelvic tumors. Int J Radiat Oncol Biol Phys 1995. In press.

— *Chapter 4* —

THE DIGITALLY RECONSTRUCTED RADIOGRAPH

Indra J. Das, PhD, Kiaran P. McGee, MS, and Gregory E. Desobry, PhD

INTRODUCTION

Historically, two modalities have been used to simulate and plan a radiation treatment. The first is a diagnostic x-ray obtained from a treatment simulator (an x-ray machine that simulates a treatment machine) and the second is a set of axial CT slices obtained from a CT scanner. A standard simulation x-ray film provides a beam's eye projection view (BEV) of the treatment portal but does not provide three dimensional information about any anatomical structures. On the other hand, CT provides anatomical information as well as target definition in each slice but does not directly provide a method to correlate this with the treatment portal. One can view anatomical structures and target volume from CT slices on a simulator radiograph by manually transferring the projection of structures in the beam direction from each slice with the proper magnification, but in general, this process is cumbersome, slow, and inaccurate. A unique approach to solving these problems has been to use contours from CT slices, projected onto a plane to digitally reproduce the conventional simulator radiograph. This has resulted in the development of the now well known digitally reconstructed radiographs (DRRs).

Like a simulation radiograph, which is the output from a conventional simulation, a DRR is essentially the output from a CT-based simulation, computed using virtual simulation software residing either on a workstation attached to a CT simulator (CT sim) or in a treatment **planning system (TPS). The DRR is a computer-generated radiograph of a virtual patient from CT data of an actual patient.** The first such image for use in radiation oncology was created by Goitein et al.[1] in 1983. That image was described simply as a projection through the CT data, and was referred to as an x-ray-like view taken from any point in space. Like conventional radiograph a DRR has built-in divergence.

METHODS OF DRR GENERATION

Review of Methodology

The development of DRRs can be traced back to as early as 1983 when Goitein et al.[1] published the first attempt to reconstruct digital projection radiographs. Cheng et al.[2] also showed that, using coordinate transformations, CT data could be used to generate "projection" images. The basic approach to creating a DRR is shown schematically in Figure 1 and typically involves four steps: ray tracing, interpolation, summation (line integration), and gray scale mapping. Using a 3D CT data set (e.g., multiple stacked CT slices), the position in space of an imaginary or "virtual" radiation source is selected. Ray lines are then cast from the virtual radiation source to an image plane perpendicular to the central axis from the source. By moving the plane up and down the central axis, the magnification or source-to-film distance is varied. Once rays are cast through the data set, the CT value along each ray line as it passes through each volume element (voxel) is calculated by an interpolation method and when the ray has passed through all the voxels, they are then summed to give an accumulated value. The intersection of the ray line and the image plane defines the pixel location on the DRR and the value of the pixel is calculated by mapping the accumulated value into a gray scale value. The image size depends on the projected size of the selected CT volume. The number of pixels in a DRR can vary typically between 80×80 and 512×512. The choice of pixel size is dependent on the required resolution and time taken to calculate the DRR. For interactive purposes a smaller number of pixels (80×80 to 128×128) may be preferred because of the significant time savings. The pixel size depends on the number of rays cast and the spacing. For a 25×25 cm^2 field with 512×512 pixels the size will be 0.5×0.5 mm^2 (250 mm/512 = 0.5mm/pixel). The pixel size for 256×256 lines will be 1×1 mm^2.

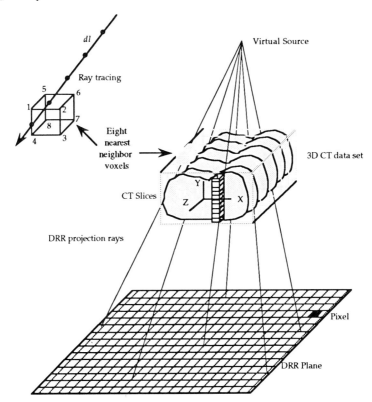

Figure 1. A schematic diagram describing the DRR generation process. The various components shown are: ray tracing through stacked CT slices; ray line interpolation among eight nearest CT voxels; summation of ray line values; projection onto the DRR image plane. For the DRR image, the number of pixels is equal to the number of rays cast. Only a few of the rays are shown.

Ray tracing is the heart of the DRR just as it is for 3D visualization.[3,4] The number of rays could be more than the number of pixels in the image plane but the number of cast rays is most frequently the same as the number of pixels in the image plane. There are two commonly used ways to cast rays depending upon the ease of computation and computer memory availability. The first involves casting ray lines such that each ray line is equally spaced on the image plane. The intersection of the rays and image plane define the pixel positions. Secondly, rays can be drawn by connecting the virtual source to the center of each pixel in the image plane. These divergent ray lines pass through the 3D CT data set and are truly representative of the ray passing from a virtual source (see Figure 1). The ray lines intersect the CT voxels at various positions. The projected pixel value in the image plane for each of these rays is the product of attenuation through each voxel cut by the ray. In a crude sense the value in each voxel is represented by the attenuation coefficient, μ, derived from the CT number of the medium for the average energy of the x-rays[5]:

$$CT = 1000\left(\frac{\mu - \mu_w}{\mu_w}\right) \quad (1)$$

where CT is the CT number in Hounsfield units (HU), μ and μ_w are the linear attenuation coefficients of the medium and water, respectively, for the x-ray energy. Eq. (1) is used to determine μ, which is then used directly in the ray line sum to compute the exponential attenuation A of the (virtual) beam as[5]:

$$A = \exp(-\sum_i \mu_i \, dl_i) \quad (2)$$

The elemental distance dl_i is the (constant) step size (Figure 1) along the ray, and μ_i is the computed attenuation coefficient in element dl_i. The step size is typically the size of a CT voxel. Step size choice is the deciding factor for the sampling rate, directly related to the time of computation of a DRR. Eq. (2) is very similar to the attenuation equation of a photon through a medium to produce a radiograph.

The value of μ_i in a step length dl_i could be computed by various interpolation methods depending upon the desired speed and accuracy. The step length is constant, governed by the desired sampling rate. Various interpolation methods have been suggested. The methods of nearest neighbor (the closest single neighbor), straight average (two neighbors), bilinear interpolation among four neighbors, and trilinear interpolation among eight neighbors are the main choices. The positions of various neighbors of a voxel are shown in Figure 1. Sherouse et al.[5] have given a detailed discussion of these interpolation methods. Trilinear interpolation requires interpolation along each axis (x, y, z) to the midpoint of the ray segment and is most commonly used for greater accuracy and resolution.[1,5,6] It was shown that trilinear interpolation is clearly superior and that other interpolation methods are undesirable.[5] The number of trilinear interpolations for a DRR is proportional to the number of DRR pixels (rays) times the number of CT pixels in one direction. There have been several attempts[7,8] to reduce this time by various ray tracing algorithms. The science of fast computer graphics and 3D surface and volume rendering use ray tracing algorithms and they are the subject of active research.[3,4] A comparison of various line integral algorithms used in ray tracing has been presented by Mitchell and Dickof,[6] who showed that with proper modifications, a speedup by a factor of ten can be achieved.

Siddon[8] used a different ray tracing algorithm in which the exact radiological pathlength through the 3D CT data set is computed based on the assumption that the density in each voxel is constant. He computed the intersections of a ray with the three sets of orthogonal planes separating CT voxels, then merged them to form an indexed set of CT voxel wall crossings along the ray. This is a fast calculation, shown to be proportional to the number of rays times the number of CT voxels along one direction.

The value of μ in each voxel, provided by the CT number, is a function of the predominant physical interactions, mainly photoelectric or Compton scattering. It is possible to selectively modify (remap) the μ values in each voxel before integration. This is usually done by look-up tables (LUTs). Based on LUTs one can weight the value of μ and create an image based on photoelectric or Compton interactions to mimic either a simulator film or a megavoltage portal film.[5] Sherouse et al.[5] showed that it was possible to create DRR images based on photoelectric, Compton, or a weighted sum image that was suggested earlier by Goitein et al.[1] One might, however, question the usefulness of creating a relatively inferior image heavily weighted by Compton scatter to match a megavoltage port film.

The weighting options to form a photoelectric or Compton image may not be directly available in some commercial systems. Most often an interactive scale is available through the choice of optimized viewing parameters for specific sites such as lung, bone, bladder, scalp, soft tissue, etc. One might also choose the LUT interactively by varying the window and level of the image. Whenever weighting options are used, an image is created by giving a weight, w_i, to the μ-value of each voxel based on the interaction. When this is done, the exponent of the attenuation A in Eq. (2) becomes

$$RS = \sum_i w_i \mu_i dl_i \tag{3}$$

The final step is to convert the ray line sum, RS, into a gray scale value for the output image. Usually RS is mapped onto an eight-bit (256 gray level) scale. The final image is a DRR which could be viewed either on a screen or a hard copy from a laser printer (paper) or a multi-format camera (film).

Computation Time

A DRR is computationally very intensive. Sherouse et al.[5] showed that on a 1 to 2 MIPS (million instructions per second) work-station, a 512×512 pixel DRR image from 50 CT slices each of 256×256 pixels takes 6 hours using the nearest neighbor calculation and 10 hours using trilinear interpolation. On a 20 MIPS workstation a DRR takes 20 minutes. A simple calculation can show how many operations are needed to compute a DRR. To read 100 CT slices of 512×512 pixels with 256 grey scales requires 6.7×10^9 ($512 \times 512 \times 100 \times 256$) operations. If a 20 MIPS processor with enough memory ($0.5 \times 100 = 50$ Mbytes) is used, then the reading time will be 6 minutes ($6.7 \times 10^9/20 \times 10^6 = 335$ sec). The ray tracing requires an additional time that depends on the orientation of the beam with respect to the CT stack. Dong et al.[7] presented a comparative DRR benchmark time study for various algorithms and ray tracing modifications. It was shown that even on a Sun SPARC station 10, a 256×256 DRR from 48 CT slices took 3.31 min, which included 1.28 min for disk reading and 2.03 minutes for ray tracing and line-integral calculation. It is clear that for a high resolution DRR these times are clinically unrealistic for interactive use. Some of the older units do not even have enough memory to load an entire CT data set. In such situations memory swapping is required, and this significantly slows down the DRR reconstruction.

Before DRR calculation can be considered clinically acceptable, the efficiency of computing a DRR must improve until it is nearly interactive. Cullip et al.[9] realized the difficulty of time saving and suggested an alternative to DRR called digitally reconstructed fluoroscopy (DRF) for interactive viewing. A DRF is simply a faster but much coarser DRR. A DRR can be computed more quickly by faster processors such as MIMD (multiple instructions and multiple data) graphic stations. Further time savings can be achieved by using bigger step sizes (i.e., by calculating ray line values in even numbered steps only) while interpolation time can be decreased by using a bilinear or the nearest neighbor method. Ray tracing density (number of rays) can also be reduced from a 512×512 to 256×256 to increase calculation speed. In some situations voxels are empty due to air and do not contribute to the ray sum. In a volume of interest (VOI), a ray line is basically not attenuated in many of the voxels. In a 3D data set a large number of voxels represent air outside the patient. By properly choosing the VOI or processing using a flag to eliminate the voxels that do not contribute to *RS*, significant gains in time can be achieved in ray tracing and line integrals. Finally, performing integer rather than floating point calculations can also realize time savings. By taking the above steps, thousand-fold savings in computation time can be achieved. In such situations, a DRR can be computed in a few seconds and used interactively. Various other modifications to make DRRs interactive have been suggested.[9,10]

The ACQSIM Method of DRR Generation

As shown earlier, a typical full quality DRR may take 5 to 30 minutes for reconstruction due to the amount of data and computations required. The system in use at Fox Chase allows full resolution images to be generated in under 20 seconds and truly interactive views at a lower resolution in under half a second. It uses trilinear interpolation at evenly spaced ray points,[1,5] but utilizes an advanced computer system. The system makes use of special-purpose rendering hardware to perform 10 MTRIPS (million trilinear interpolations per second). For a processor having 10 MTRIPS, the theoretical time for tracing 512×512 rays through 512 CT pixels per ray is 13.4 seconds ($512 \times 512 \times 512/10 \times 10^6$ sec).

A highlighted overview of the system is shown in Figure 2. The memory in the workstation is normally adequate to load the entire 3D CT data set (maximum of 256, 512×512 pixel images), which eliminates memory swapping. Rays are cast from a point corresponding to the radiation source. The maximum number of rays in the ACQSIM system is 900×900, but fewer rays (usually 512×512) are used for a clinical DRR. These rays pass through the CT

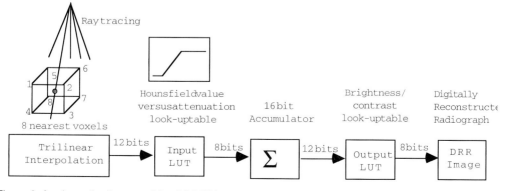

Figure 2. A schematic diagram of the ACQSIM process for generation of DRRs.

image data set to the desired image plane. The sampling of the CT data in each ray is uniform along the ray. The smallest sampling size is the size of the pixel in the CT image. For interactive viewing, the step size is relatively large (two or four pixels long). At each sampling point, trilinear interpolation among the eight neighboring voxels is used (see Figures 1 and 2). In this highly optimized implementation there is no option to use a bilinear or nearest neighbor approach. The trilinear interpolation is very effectively integrated into the ray tracing process by the pipelined hardware architecture of the special purpose workstation.

At each step along a ray, the interpolated value of the CT number is mapped into one of four dynamically allocated look-up table (LUTs) functions provided in the system. The four functional forms of these input LUTs are shown in Figure 3. The 12-bit information from the CT number is condensed to eight bits after LUT mapping, and accrued in a 16-bit accumulator. There is a possibility of overflow in the accumulator, and this is checked at each point. After a ray has been summed the accumulator contents are stored in a special 12-bit buffer. Once all the rays are cast and processed, the output data from the individual 12-bit accumulated values stored in the special purpose buffer memory are normalized to eight-bit values for display purposes. This normalization through an output is provided via user controlled brightness and contrast adjustments. The eight-bit values provide 256 gray scale values in the final DRR image.

The number of rays cast is simply a function of the output image screen: one ray goes to each pixel in the image plane. To achieve the desired interactivity, so that the image is updated in near real time as the viewing angles are adjusted, a coarse-to-fine strategy similar to that suggested by Cullip et al.[9] is employed. This involves casting fewer rays initially, and after user interaction is suspended casting the full number of rays to produce the final image. The coarse view is sub-sampled by a factor of 16 from the final, but the image quality is enhanced by simple bilinear interpolation of the output pixels resulting in an acceptable image for interactive work. In addition, a VOI may be selected so as to reduce the number of sampling points

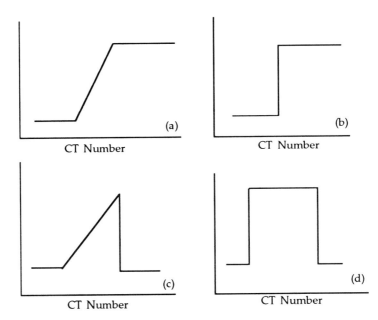

Figure 3. Functional forms of the input look-up table (LUT) used in preprocessing: (a) ramp function, (b) step function, (c) shark function, and (d) boxcar function.

along each ray. For example, a lateral radiograph of a pelvis involves casting rays through the outer layers of muscle, fat, and air. The computation time is significantly reduced by using a smaller VOI. Application of the smaller VOI has other advantages too. The time to load the image data from disk to the workstation memory is reduced and the quality of the image is enhanced by the removal of attenuating structures. The computation time on the system for the final DRR image varies from about 5 seconds for a 512×512 image with a 50% VOI to about 20 seconds with no VOI applied.

It should be pointed out that the system does not support or attempt to account for scattering of photons in this model, and hence images based on physical interaction probabilities (photoelectric and Compton) are not directly possible. It was indicated by the manufacturer that their earlier test results showed the attenuation and scattering model did not provide any intrinsic advantage over the simple linear LUTs. In fact, adjustment of the input LUTs provided greater flexibility in determining the output contrast and brightness of the resulting DRRs.

Although there is no explicit energy selection built into the DRR generation process, DRRs of differing contrast may be produced by interactive adjustment of the both the input and output LUT mapping functions. The combination of these functions can in fact be used to mimic images produced by a variety of differing imaging modalities and energies from low (80 to 150 kVp) through megavoltage beams. For example, Figures 4a and 4b show photoelectric- and Compton-weighted like images, respectively, of the thorax of a patient. These images are generated to enhance the bony and soft tissue structures. In reality, the images shown in Figure 4 are produced by the system using prestored windowing and level LUTs for bone and soft tissue.

IMAGE QUALITY

Image quality is a critical issue in terms of the clinical acceptance of DRRs in routine radiotherapy practice. Unlike simulation film images which are theoretically limited by the grain size of the film, DRR resolution is limited by the pixel size of the CT slices, their thickness and separation. The difference between these can be many orders of magnitude, as simulation film grain size is typically one micron whereas CT pixel dimensions range between 0.5 and 1.0 mm. Another factor that degrades image resolution in both simulation and DRR images is the effect of overlying structures. In simulation images, absorption and scattering affect target volume resolution. Preferential absorption by materials with high Z numbers, such as a prosthesis, can cast a "shadow" over other structures on the simulation film while the scattering of incident x-ray photons either proximally or distally to the target volume serve to decrease the contrast of the target on the simulation image. The advantage of DRRs in this respect is that the effect of absorption on target volume resolution can be reduced, and scattering (except that inherent in the CT images) is eliminated. The mere act of taking axial CT slices reduces scattered photons, as only a thin slice of the patient is imaged at a single time. The effective absorption of photons in overlying tissues can also be reduced by using a wireframe or solid shading of the target as suggested by Weinhous et al.[10] or by selecting a volume of interest that encompasses the target and some other critical structures for treatment planning and verification purposes. Several other, secondary factors can also decrease the image quality of DRRs. McGee et al.[11] have outlined several of these and have divided them into two groups based on the types of errors that they can cause. Specifically these are reconstruction

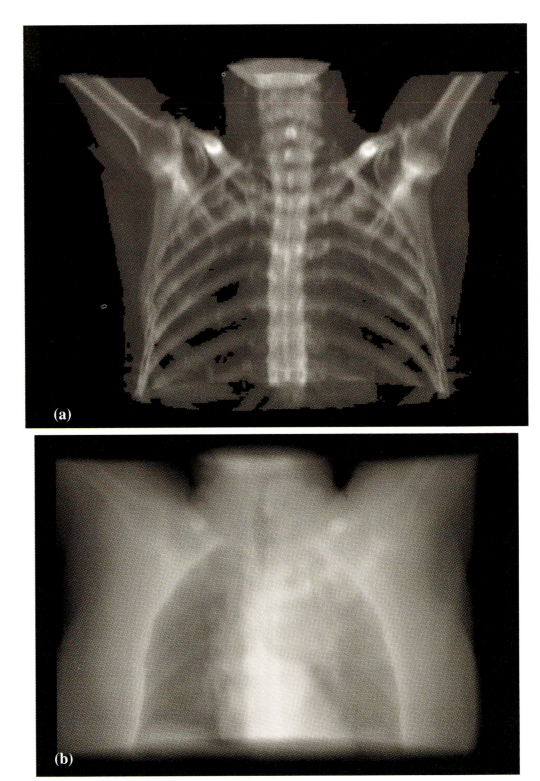

Figure 4. DRR images of the thorax of a patient, representing photoelectric- and Compton-weighted images for (a) bone and (b) soft tissue, respectively. The images can be changed interactively by changing the dynamic range of the LUT to see a specific organ which may not be seen otherwise.

registration errors and ray line divergence errors.

Finally, DRR image quality is also a function of CT slice thickness and separation between slices; voxel size or the size of the reconstruction circle used to acquire the CT slices, and the DRR reconstruction algorithm. In general, smaller slice thicknesses, separations, and voxel sizes provide a better DRR image. Gilhuijs et al.[12] have demonstrated that the appropriate choice of algorithm can reduce effects such as partial volume averaging, especially at bone/soft tissue interfaces. As the thickness of the axial CT data has the smallest resolving power, it is not surprising that this parameter determines the resolution (sharpness) of the DRR. Figure 5 is a demonstration of the effect of slice thickness on resolution. Figure 5a is an anterior DRR of a humanoid CT phantom where the slice thickness and separation are 5 mm and 5 mm respectively. Figure 5b shows the same phantom and scan volume except the slice thickness and separation are 2 mm and 2 mm, respectively. The superiority of the image with smaller slice thickness is clearly visible.

USES OF DRRS

Substitute for a Simulator Radiograph

DRR images mimic the properties of a conventional x-ray film and can be used to aid in the localization and verification process of radiation therapy simulation prior to radiation treatment. The most important use of the DRR is as a substitute for a conventional simulation radiograph. It can also be used to visualize the target volume and other structures by BEV overlay of contours of these structures previously outlined on CT slices. DRRs can be created with different window and level setting options. They can emphasize a bony structure or a soft tissue structure whose discrimination may prove useful in clinical practice. Unlike simulator radiographs, which cannot be changed, DRRs can be viewed and changed interactively with dynamically adjusted LUTs to visualize a desired structure. An example of this is seen in Figures 4a and 4b in which bony and soft tissue structures are visualized by selective enhancement.

Digital Image for Processing

Since a DRR is a digital image it has the advantage that it can be electronically manipulated. A typical example is the change of contrast to view structures for clinical application. By various other image enhancement techniques it is possible to see some structures which would otherwise be missed. A digital image can be integrated into a departmental database and is easily portable through a network. It is easier to archive images in a picture archiving and communication systems if they are digital in nature. A detailed description of such a system was recently published by McGee et al.[13] DRRs are also gaining popularity in other fields. A recent application of a DRR for image registration in radiosurgery has been developed.[14]

Treatment Verification

A DRR can be compared to a portal image acquired at the linear accelerator to determine set up accuracy and detect patient movement. Electronic portal imaging devices (EPIDs) are replacing the traditional megavoltage portal films. DRR images can be used as a gold standard

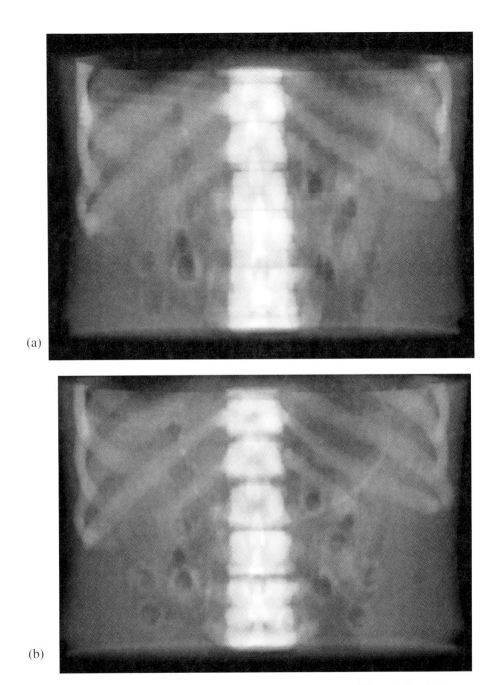

Figure 5. DRR images of the section of abdomen of a CT Phantom. Effect of slice thickness and table increment is visible. (a) 5 x 5: 5 mm slice and 5 mm table increment image, and (b) 2 x 2: 2 mm slice and 2 mm table increment image. Note that image (b) has greater resolution than image (a).

for treatment verification against either EPIDs or digitized portal films once their accuracy is verified. Verification of the geometrical accuracy of a DRR system was discussed in a recent paper.[15] An overview of DRR quality assurance can be found in Chapter 2. DRR images can be used for treatment verification qualitatively.

FUTURE TRENDS

DRRs may be accepted as a substitute for simulator radiographs. Acceptance will depend upon their quality and the time required to compute them. The quality of DRR images may improve significantly with the development of faster processors and better functional forms. Picker is exploring the use of weighted non-linear mapping functions instead of those shown in Figure 3 for input processing to selectively enhance some range of CT values with respect to other values. There are promising results in soft tissue visualization where certain soft tissue types can be highlighted in a projection view, giving a better, selective DRR. Another new trend in visualization, called digital composite reconstruction, is being explored. This allows the visualization of structures that have different CT values. This would be done on the basis of iso-CT mapping in a selected VOI. Volume rendering and later projection is done in a way similar to current DRRs. These images may provide automation in structure delineation without contouring the structure in each slice.

Faster processors and ray tracing algorithms are being developed to make the DRR interactive. The ACQSIM method of coarse to fine sampling is a single step refinement. The use of a pyramid sampling approach, also known as successive refinement, could also be used. This might help interactivity in some cases. Certain filtering processes not yet used may have a role in improving the image quality of DRRs from thicker CT slices.

SUMMARY

The DRR is a computer-generated radiograph formed by projecting ray lines through a set of axial CT scans of a patient. It has the same features as those of a simulator radiograph. DRRs are generated from virtual simulator software residing in a CT simulator or a treatment planning system. They are digital in nature and hence can be electronically manipulated to adjust contrast and brightness, to add annotations, to cut and paste, etc. A DRR is computationally intensive and requires a large memory and fast processing. A high-end graphics machine is desirable for generating DRRs. DRRs have enormous potential for clinical application and are probably the future of radiation oncology for simulation and verification.

REFERENCES

1. Goitein M, Abrams M. Multidimensional treatment planning: II. beam's eye-view, back projection, and projection throughout CT sections. Int J Radiat Oncol Biol Phys 1983;9:789-797.

2. Cheng CW, Chin LM, Kuewski PK. A coordinate transfer of anatomical information from CT to treatment simulation. Int J Radiat Oncol Biol Phys 1987;13:1559-1569.

3. Udupa JK. Visualization of Images. Madison, Wisconsin: Medical Physics Publishing, 1993.

4. Herman GT. Image Reconstruction from Projections: The Fundamentals of Computerized Tomography. New York: Academic Press, 1980.

5. Sherouse GW, Novins K, Chaney E. Computation of digitally reconstructed radiography for use in radiotherapy treatment design. Int J Radiat Oncol Biol Phys 1990;18:651-658.

6. Mitchell JR, Dickof P. A comparison of line integral algorithms. Computers in Physics 1990; 4:166-172.

7. Dong L, Kachilla D, Boyer AL. A ray tracing program based on Bresenham algorithm for generating digitally reconstructed radiographs. Med Phys 1994;21:886.

8. Siddon RL. Fast calculation of exact radiological path for a three-dimensional CT array. Med Phys 1985;12:252-255.

9. Cullip TJ, Symon JR, Rosenman JG, Chaney EL. Digitally reconstructed fluoroscopy and other interactive volume visualizations in 3D treatment planning. Int J Radiat Oncol Biol Phys 1993;27:145-151.

10. Weinhous MS, Li Z, Holman M. The selection of portal aperture using interactively displayed beam's eye sections. Int J Radiat Oncol Biol Phys 1991;22:1089-1092.

11. McGee KP, Das IJ, Schultheiss TE, Sims C. A quality assurance phantom for digitally reconstructed radiographs (DRRs). Med Phys 1994;21:902.

12. Gilhuijs KGA, Drukker K, van de Ven PJH, van Herk MB. A new method for fast generation of digitally reconstructed radiographs (DRRs). Proceedings of the XIth International Conference on the Use of Computers on Radiation Therapy, Manchester, 1994, 228-229.

13. McGee KP, Das IJ, Fein DA, Martin EE, Schultheiss TE, Hanks GE. Picture archiving and communications systems in radiation oncology: tools for a physician-based digital image review system. Radiother Oncol 1995;34:54-62.

14. Lemieux L, Jagoe R, Fish DR, Kitchen ND, Thomas DGT. A patient-to-computed tomography image registration method based on digitally reconstructed radiographs. Med Phys 1994; 21:1749-1760.

Chapter 5

CT SIMULATION HARDWARE

Eric E. Martin

INTRODUCTION

Simulation provides for visualization of the treatment field, marking the patient for treatment, and treatment field verification images. Conventional simulators use an image intensifier to provide real-time viewing of treatment fields with diagnostic energy photons (fluoroscopy). Patient marking is provided by at least two lateral stationary lasers arranged to delineate the machine isocenter. In addition, cross-wires in the light field of the simulator header locate the isocenter of the treatment field. If the simulator provides for block inserts the light field can also be used to mark the field outline on the patient. Conventional simulators produce treatment field verification images by exposing radiographic film on top of the image intensifier. As treatments and dose calculations become more complex the use of CT scans of the treatment area is becoming more common. The goal of CT simulation is then to replace the functionality of the conventional simulator through the use of the CT data gathered in the treatment planning CT scan.

The basic operations necessary to perform CT simulation are the production of a CT study, virtual simulation mimicking to some extent fluoroscopy as performed on a conventional simulator, and the production of digitally reconstructed radiographs (DRRs). In addition, CT simulation often provides for localization marking of the patient. Several configurations of hardware are possible which can be classified as CT simulation. Any discussion of specifics about the hardware required for CT simulation must be predicated on an understanding of the various possible configurations. Figures 1 through 3 illustrate several possible hardware configurations.

The configurations presented in Figures 1 through 3 cover a wide variety of approaches to CT simulation. However, while some hardware specifics may differ with various implementations, the basic capabilities and requirements of the hardware remain constant.

CT SCANNER

Many CT scanners available in the market today are capable of providing CT studies appropriate for CT simulation. There are, however, several features which can enhance a CT scanner's applicability to CT simulation. One feature of importance is the gantry aperture. Most CT scanners have a 70 cm aperture which is adequate for the majority of patient set ups.

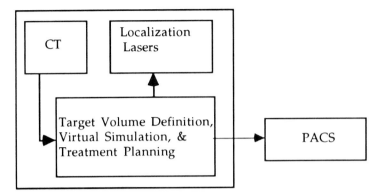

Figure 1. Vendor equipment performs CT scan, the CT study is transferred to a workstation where target volume definition is performed. Isocenter location is then transferred to localization lasers. Virtual simulation is performed on the workstation producing DRRs. RTP software on the workstation allows a treatment plan to be produced. Optionally, DRRs from the virtual simulation can be transferred to a PACS. (A system providing this high level of integration is not yet currently available.)

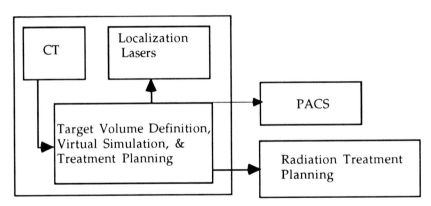

Figure 2. Vendor equipment performs CT scan, the CT study is transferred to a workstation where target volume definition is performed. Isocenter location is then transferred to localization lasers. Virtual simulation is performed on the workstation producing DRRs. CT slices, contours, and field information is transferred to an external treatment planning system. Optionally, DRRs from the virtual simulation can be transferred to a PACS.

Apertures smaller than 70 cm would greatly limit the variety of patient set ups which could be performed while larger gantry apertures would increase the ability of the scanner to accommodate larger patients in an assortment of set up positions. Ideally the aperture would be 80 cm or greater, however, as the size of the aperture increases the number of detectors, intensity of radiation (to compensate for greater source-to-object distance), and travel distance of the gantry components also increase. Thus the gantry aperture size is limited by cost factors, increase in patient dose, and increase in scan time.

Variable slice thickness and spacing are extremely valuable in obtaining CT studies conducive to producing high quality DRRs. Many scanners allow only certain useful slice thickness settings (e.g., 2 mm or 5 mm) and slice spacing (e.g., 2 mm, 3 mm, or 5 mm) during

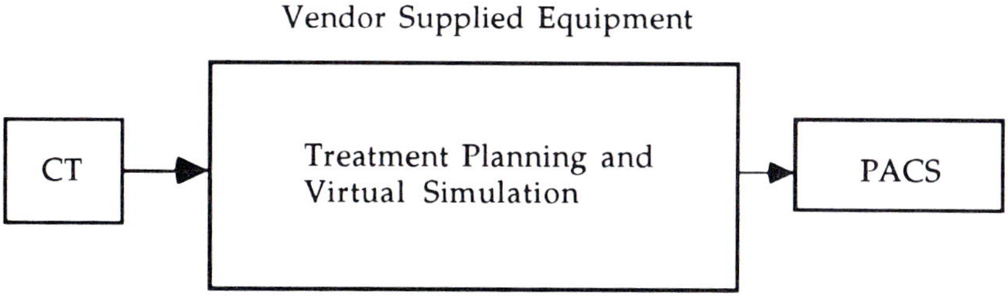

Vendor Supplied Equipment

CT → Treatment Planning and Virtual Simulation → PACS

Figure 3. Vendor equipment imports the CT study from a CT scanner and allows treatment planning to be performed. Virtual simulation can be performed to assist in the planning process. DRRs are produced of the final treatment fields. Optionally, DRRs from the virtual simulation can be transferred to a PACS. This configuration must be accompanied by conventional simulation for patient marking.

acquisition. Fixed settings can severely hamper efforts to increase transaxial resolution at critical anatomical structures. For example, when performing CT simulation on a head and neck case acquiring 1 mm slices at 1 mm spacing through the region containing the tumor and 3 mm slices at 3 mm spacing through peripheral areas maximizes the useful resolution of the resulting DRRs while minimizing the impact on storage and communication of the CT study.

SPIRAL CT SCANNERS

Spiral CT scanners are especially applicable to CT simulation. Spiral scanners transport the patient through the gantry aperture while continuously rotating the x-ray source and acquiring data. The resulting CT data resembles a slightly uncoiled spring and requires special processing to reconstruct axial CT slices. The major advantages of spiral scanning are: reduction in acquisition time (because of the elimination of the interscan delay time), patient motion is reduced (thus organs and other tissue can be more accurately delineated), and slices can be obtained for an arbitrary position in the scanned volume. The rapid acquisition of spiral CT in combination with timed contrast administration offers many novel uses of CT for diagnostic purposes. Oncologic uses of spiral scanners require more tube performance than typical diag-

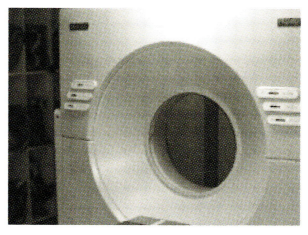

Figure 4. CT Scanner with 70 cm gantry aperture showing patient couch and gantry controls.

nostic uses, since larger volumes are generally scanned and smaller table movement per revolutions are used to increase transaxial resolution. Tube loading is hence much higher and tubes which can handle heavy loading are invaluable to the process.

The ability of a spiral scanner to decrease the time necessary to acquire a full CT data set is valuable in CT simulation. In addition, the ability to reprocess the CT data to produce axial slices with different thickness and spacing parameters can be extremely useful in a clinical setting. After the initial acquisition of a CT data set, the data set is processed to produce 1 mm slices spaced at 5 mm. These slices are then transferred to the target volume definition station where the isocenter of the target volume is selected, the coordinates transferred to the localization lasers, and the isocenter markings placed on the patient. Later, after the patient has been marked and gone home, additional slices are produced from the CT data set as needed to increase the resolution of the final treatment field DRRs.

CT SIMULATION TIME CONSIDERATIONS

Regardless of whether an axial or spiral CT scanner is used, the speed of CT slice reconstruction and CT slice transfer to the target volume definition station is critical if patient marking of the isocenter or treatment fields is to be performed. The major clinical advantage of patient marking during CT simulation (as compared to orthogonal marking during CT acquisition and subsequent isocenter marking or working with an offset) is removing a patient set up and thus a possible set up error. However, the time between the start of acquisition and patient marking must be minimized to counter the increasing probability of patient movement with increasing time on the CT couch. Minimizing the acquisition and reconstruction time is an important step in decreasing the overall time that the patient is required to remain motionless on the CT couch.

Another important step in reducing patient waiting time is minimizing the time required to transfer CT slices to the target volume definition station. Commonly, an Ethernet network connection between the CT scanner and the target volume definition station provides the communications link. Increasing the speed of this network connection by utilizing a fiber optic FDDI or a Fast Ethernet link increases the transfer speed. In addition, the capability to transfer slices to the target volume definition station as they are reconstructed allows the definition of the target volume to commence prior to the receipt of the entire CT study. The simultaneous transfer of slices during acquisition not only decreases the importance of the transfer speed, but also allows the times required to acquire the CT study and delineate the target volume to be overlapped. Table 1 delineates the average times required to complete the various components of CT simulation which require the patient to remain motionless on the couch.

CT CONSOLE

The CT console is a computer system which controls the CT scanner, performs the reconstruction of the CT data, provides for storage of the CT slices, allows manipulation and viewing of the CT slices, and communicates the CT studies to external devices (such as a virtual simulation workstation). The capabilities and design of the CT console greatly influence the efficacy of the CT scanner.

Memory requirements for a CT console differ for axial and spiral scanners. Most axial CT

CT SIMULATION HARDWARE

Table 1. Typical CT Simulation Times Requiring Patient to Remain Motionless (60 Slice Study)

	Axial CT	Spiral CT	Simultaneous Transfer
Acquisition and reconstruction	15 minutes	10 minutes	10 minutes
Transfer of slices	5 minutes	5 minutes	negligible
Target volume definition	10 minutes	10 minutes	10 minutes
Total time patient is motionless	30 minutes	25 minutes	12 minutes*

*Assumes target delineation is completed within 2 minutes following transfer of last required CT slice.

scanner consoles deal with a CT study one slice at a time even during reconstruction. Hence memory of 16 to 32 megabytes is generally sufficient for an axial CT scanner console. Some axial scanners use a distributed processing bus design which allows reconstruction hardware to function at the same time additional slice data are acquired. Consoles of these scanners often contain up to 64 megabytes of memory to allow for buffered data acquisition. While performing slice reconstruction a spiral CT scanner console must access the raw CT data set repeatedly. Optimally the console would be able to keep the entire raw data in its memory during the reconstruction process. However, selective portions of the raw data can be retrieved from a disk as needed. Generally 64 to 128 megabytes of memory provides for sufficient performance in a spiral CT scanner console.

Storage requirements for a CT console also differ for axial and spiral scanners. A spiral scanner console should have the capacity to store several raw data sets in addition to the reconstructed CT slices. By storing the raw data set, additional CT slices can be produced by varying the reconstruction parameters. The raw data set can require from 300 to 500 megabytes of storage per study depending on the scan time and couch translation speed (*calculation for raw data size:* number of samples per view [1024] × number of bytes per detector [2] × number of views per rotation [360] × number of rotations [700 (1 mm per rotation for 70 cm)]) / 1024^2 bytes per megabyte = megabytes of raw data per typical study [492 megabytes]). Thus this storage space must be added for two to three studies when a spiral CT scanner is being considered. A typical per study storage budget for reconstructed CT slices allocates .5 megabytes per slice. CT studies performed for CT simulation often contain more slices per study than CT studies performed for diagnostic purposes (because of the need to increase transaxial resolution to produce high quality DRRs). Hence an average of 70 slices per study is not unreasonable. Each study would therefore require approximately 35 megabytes of storage. For a spiral scanner 35 megabytes of storage for the reconstructed CT slices plus approximately 500 megabytes of storage for the raw CT data set or approximately 535 megabytes total is needed. Since retrieval from the CT scanner during the course of a work week is probably reasonable and assuming that a full load of cases is eight per day, the storage budget for axial and spiral CT scanners is as shown in Table 2.

Hence, using these conservative figures an axial CT scanner console should have access to approximately 1.5 gigabytes of local storage space for CT information and a similarly sized tape backup mechanism. A spiral CT scanner console should have access to approximately 3 gigabytes of local storage space for CT information and a similarly sized tape backup mechanism.

Table 2. Storage Budget for CT Scanner Console Used for CT Simulation*

Scanner Type	Days	Patients per Day	Slices per Patient	Megs per Slice	Raw Data Sets	Total Storage
Axial	5	8	70	.5	NA	1.5 GB
Spiral	5	8	70	.5	1.5 GB	3 GB

*Allowing for storage of three raw data sets.

PATIENT MARKING

Patient marking can be an important function of CT simulation. Existing commercial systems allow for either isocenter or field edge marking. Prior to being scanned the patient is temporarily marked orthogonally to provide verification that the patient did not move during the scanning process. Isocenter marking is performed after target volume delineation. An isocenter is chosen and the contouring package calculates the proper laser positions marking the isocenter points on the patient. If the longitudinal motion of the couch is sufficiently accurate (<1 mm) stationary lateral laser can be used and the couch moved to indicate the isocenter. To delineate the lateral isocenter position a movable laser can be used.

VIRTUAL SIMULATION WORKSTATION

Depending on the design of the CT simulation system, the virtual simulation workstation may include software to perform target volume definition and/or treatment planning and dose calculation in addition to virtual simulation and DRR production. In any case the hardware required by the virtual simulation workstation can be defined by the operation of producing DRRs. The production of good quality DRRs requires massive amounts of memory and processing power. DRRs are produced by tracing the divergent path of simulated photons from a point source through CT data and recording the attenuation encountered by the "photon" using the relationship between CT numbers, the electron density of the material, and photon attenuation (see Chapter 4 for a complete discussion of DRR generation).

The most efficient method to generate DRRs is to load all of the CT slices in a study into memory so that the ray tracing can be performed without swapping information from its storage. For an average 70 slice study this would require at least 35 megabytes of memory. Some virtual simulator workstations contain at least this much memory. However, disk swapping can be used to reduce the memory requirements. If disk swapping is used the time to produce a DRR can be increased by a factor of 3 or 4 depending on the efficiency of the swapping algorithm. Ray tracing requires that the processor frequently access large portions of its memory, therefore, the speed at which the processor can access its memory is critical. For example, a processor with an external bus speed of 50 MHz and an internal clock speed of 100 MHz would produce a DRR more slowly than a 75 MHz processor with a 75 MHz bus speed. In normal operations a primary and secondary cache keeps the processor with the slower external bus speed from making excessive accesses to its external memory, however, ray tracing through 35 megabytes of CT data renders the caches impotent. The virtual simulation work-

Figure 5. Movable laser unit facing the scanner gantry for patient marking.

stations of some systems contain specialized array processing hardware which greatly speeds the ray tracing process.

In order to perform virtual simulation (which approximates fluoroscopy on a conventional simulator) the production of DRRs must occur in close to real time. Currently, no system can produce full-size, high-resolution DRRs in real time. In order to approximate the full-size, high-resolution DRR ideal, systems which perform virtual simulation should provide a fast production mode where ray tracing using a coarse calculation grid is performed. The use of a coarse calculation grid produces low-resolution DRRs in close to real time. Other systems provide no virtual simulation capability, instead producing only a three dimensional beam's eye view (BEV) of anatomical structures contoured from the CT data to allow for aiming of the beam and produce DRRs of the final treatment beams. Although the production of BEV displays is much faster than the production of DRRs, less anatomic information is available in BEV displays and the time consuming operation of contouring anatomy in multiple planes is required.

Calculation times for DRRs vary widely from system to system depending on the amount of memory, processor speed, bus speed, algorithm used, and the presence of any specialized hardware. One system which contains 64 megabytes of memory and specialized rendering hardware can produce a full-size, high-resolution DRR in 4 to 10 seconds (depending on the number of slices in the study, the magnification, and other parameters). Another system using 32 megabytes of memory and no specialized rendering hardware produces a full-size, high-resolution DRR in 1 to 2 minutes.

In addition to the large amounts of memory and processing capability required, the storage needs of the virtual simulator are also considerable. The virtual simulation workstation must store the CT study, DRRs, field parameters, and patient information (if treatment planning is also performed on this station, then plan information and estimated dose grids must also be stored). The largest, and hence most important, of these items are the CT study (which on average we previously showed to consume approximately 35 megabytes of storage), DRRs (which at a resolution of 512 pixels $\times$ 512 pixels $\times$ 2 bytes consumes .5 megabytes per DRR), and the dose grid if applicable (which at $64 \times 64 \times 64 \times 4$ bytes consumes 1 megabyte per dose grid). If we assume one CT study per patient, eight DRRs per patient on average, and two dose grids per patient and wish to store one month's worth of patients operating at a full work load of eight patients per day, the storage budget for the virtual simulator workstation is as shown in Table 3.

Hence, using these figures, a virtual simulator workstation could need access to approximately 7.5 gigabytes of local storage space for CT information, DRRs, and treatment planning information (if applicable) and a similarly sized tape backup mechanism.

ADDITIONAL CONSIDERATIONS

As demonstrated above, the amount of data being handled by all of the systems involved in the CT simulation process is significant. The networking scheme used to allow the systems to transfer this amount of data must be capable of handling large sporadic loads. For example, transfer of an entire CT study (35 megabytes) over a typically loaded standard Ethernet network (10 megabits per second) can take up to five minutes. While this may be acceptable for a one-time transmission of data, it quickly becomes impractical if consistent retrieval from a central storage location is required. For this use a network topology capable of handling larger loads at faster speeds such as FDDI or Fast Ethernet should be considered.

If the virtual simulation station is used to perform target volume definition, virtual simulation, DRR production, and treatment planning a single station can easily become overused. Setting up an additional virtual simulation station should be provided for in the design of the CT simulation system. One of the virtual simulation stations can provide for storage of the CT simulation information or a dedicated file server can maintain these files. If centralized storage is used, the speed of the network connection among the virtual simulation workstations and the file server is extremely important. To provide adequate performance while routinely retrieving data sets in excess of 30 megabytes an FDDI or Fast Ethernet link of 100 megabits per second should be used.

CONCLUSIONS

Computer technology capable of performing CT simulation in reasonable time is rapidly developing. Fast Ethernet, FDDI, workstations with 64 megabytes of memory, and 9 gigabyte hard drives are recent developments. However, in a few years the hardware required for CT simulation will be common technology instead of new, expensive, and hard to find. The hardware of the future will provide CT simulation with additional capabilities, increases in speed, and at a reduced cost. However, it is almost certain that the capabilities and tasks of CT simulation will expand to consume these advances in technology.

Table 3. Storage Budget for Virtual Simulator Station

RTP	Days	Patients per Day	CT Study	DRRs	Dose Grids	Total Storage
No	23	8	35	4	NA	7 GB
Yes	23	8	35	4	2	7.5 GB

REFERENCES

1. Seeram, E. Computed Tomography: Physical Principles, Clinical Applications, and Quality Control. Philadelphia, Pennsylvania: W.B. Saunders, 1994.

2. Picker International, Inc. PQ 2000 CT Imaging System. Cleveland, Ohio, 1993, BR1122 R1193.

3. Picker International, Inc. ACQSIM Oncodiagnostic Simulation/Localization System. Cleveland, Ohio, 1993, PD1075 R1193.

4. Picker International, Inc. Voxel Q Visualization System. Cleveland, Ohio, 1994, BR1184 594.

Chapter 6

THE TREATMENT PLANNING PROCESS

Margie Hunt, MS and Lawrence Coia, MD

INTRODUCTION

Treatment planning is a complex process which begins with the decision to treat a patient with radiation and ends with verification of the treatment fields on the treatment machine. The major steps in the treatment planning process include:
- Definition of treatment intent
- Immobilization
- Localization
- Field design/graphical planning
- Treatment verification

Although these steps have always formed the basis for treatment planning, technological advances during the past twenty years have drastically changed their implementation. The increased precision, accuracy, and reproducibility achieved with this technology allow patients to be treated to higher and more uniform doses with the same or fewer normal tissue complications. Improved local control and survival are the anticipated outcomes although the link between technical aspects of treatment and outcome is not necessarily well understood. One recent study[1] has elucidated the relationship between local control and survival for prostate cancer. Other recent studies[2-4] have evaluated the relationship between precision radiation therapy, dose, and local control for prostate cancer. More such studies are essential to provide guidance for the efficient and wise use of new technology.

At each step in the treatment planning process, there are vigorous ongoing efforts to refine the application of new technology to patient treatment. Radiobiological and clinical research continuously impact on the choices of dose, fractionation schemes, chemotherapeutic, surgical and combined approaches which must be considered by the radiation oncologist as they formulate their treatment intent for each patient. The trend toward smaller, more precisely defined treatment fields has led to a heightened appreciation of the value of patient immobilization. New materials and methods have become available which allow custom devices to be made for each patient. Numerous studies of patient set up reproducibility on the treatment machine have been published for a variety of sites and methods of incorporating the effect of this uncertainty into treatment planning are appearing (Appendix I). Boyer et al[5] give an excellent review of electronic portal imaging devices and their impact on verification. In a sense, however, these technological innovations have occurred in response to the revolution begun with the use of CT in treatment planning. CT has radically changed treatment planning and treatment complexity by individualizing localization and field design for each patient. CT-assisted three dimensional treatment planning is now routine in major centers and becoming increasingly common in smaller departments as well. State-of-the-art CT planning incorporates information from a variety of imaging modalities to create a three dimensional reconstruction of the patient's normal and malignant anatomy. The CT technology explosion continues with spiral scanners and sophisticated image fusion software which melds images from multiple modalities. The radiation oncologist's dilemma is how to incorporate an increasingly bewildering amount of information into the treatment planning process.

The purpose of this chapter is to describe each step in the treatment planning process with particular emphasis on the role of CT simulation in each step. Prior to that discussion, however, a short history of the development of CT simulation is in order.

The earliest CT simulator was conceived as a tool to provide basic simulation capability, including patient marking, with treatment planning functions integrated. It consisted of a CT scanner, laser marking system with driving software and a treatment planning system. Such a device was actually marketed by Pfizer in the late 1970s and early 1980s. Due to several factors, including the lack of digitally reconstructed radiographs (DRRs), this device was not a commercial success. The integration of CT scanning with simulation was still not well defined, at least in the minds of the users, making this product difficult to incorporate into the clinic. Users welcomed the link between CT scanning and patient marking as a solution to some of the inaccuracy and inefficiency inherent in standard CT-assisted treatment planning but they were unable to justify the inclusion of a treatment planning system. A conventional simulator does not perform treatment planning per se, although it is intimately related to the treatment planning steps of localization and field design. Difficulties arose when an attempt was made to define a treatment plan as part of a simulation session, particularly for complex treatments requiring isodose calculations. The technology in the early to mid 1980s did not allow for interactive definition of target volumes or interactive dose calculation and display, hence the ability to develop a treatment plan as part of the simulation process was not practical.

The other major factor in the limited success of early CT simulation systems was the non-availability of high quality DRRs. These images have replaced conventional simulation films as the gold standard for treatment verification. Without DRRs, the patient must receive a conventional simulation, adding time and potential inaccuracies to the treatment planning process. Subsequent CT simulation systems have provided DRRs, but without the quality necessary to replace simulator films. Indeed, DRR image quality is still a major issue, and conventional simulation is still sometimes needed even with high performance systems such as the one at Fox Chase Cancer Center.

State-of-the-art CT simulators incorporate a high performance CT scanner with a localization package, patient marking system and a virtual simulator capable of producing real time DRRs for use in field design. As discussed below, the CT simulator has a major impact on the process of treatment planning particularly in the areas of localization and field design.

TREATMENT INTENT

The definition of the intended treatment is the initial step in the treatment planning process. Staging of the patient and the determination of prognosis and goals of therapy are all major issues which must be dealt with by the radiation oncologist before any technical planning can begin. Once these issues are resolved, the physician must then formulate the treatment prescription which includes a description of the target volumes and normal tissues at risk in addition to a statement of dose and fractionation schemes. Complex treatment planning such as that done with a CT simulator and a three dimensional treatment planning system requires much more than the average amount of communication between physician and planner. Our policy is to request all of the prescription information described below from the physician or resident at the time the patient is scheduled for immobilization and localization procedures. The form used at Fox Chase Cancer Center to obtain this information is shown in Appendix II.

Anatomical Description of Target Volumes

To describe the anatomical regions relevant to radiation oncology, the nomenclature defined in the ICRU 50 report[6] is used throughout this book and is briefly described below:

ICRU 50: Prescribing, Recording and Reporting Photon Beam Therapy
  Gross Tumor Volume (GTV) — Palpable or visible extent of the tumor.
  Clinical Target Volume (CTV) — GTV plus local margins for subclinical disease or regions of presumed microscopic disease.
  Planning Target Volume (PTV) — CTV plus margins for treatment reproducibility factors such as patient and organ movement, respiration and daily set up but excluding margin for beam penumbra.

As part of the prescription, the physician should include a general anatomical description of the various GTVs and CTVs which they intend to treat. For any given patient, there are often multiple GTVs and CTVs describing regions which are anatomically disjointed or which have different probabilities of disease. As an example, a patient being treated for nasopharynx cancer may have at least two GTVs; the primary disease and a grossly involved lymph node in the neck. The CTVs for the same patient would include each of the GTVs with a margin for subclinical extension in addition to nodal regions such as the posterior cervical and supraclavicular chains which are treated bilaterally because of presumed microscopic disease. Since the GTV and CTV are anatomical concepts, they are not related to the actual treatment fields and therefore can be defined prior to any field definitions.

These target volumes are given well-defined contours yet each volume is in fact probabilistically determined. The GTV is likely to represent the actual gross tumor volume, however, one must consider inaccuracies in defining gross tumor with CT simulation. Thus, the GTV representation of actual gross tumor volume depends on a number of parameters including the tumor site, scanning parameters such as slice thickness and slice separation, and the skill of the physician in interpreting the CT scan and defining tumor versus normal tissues. Since microscopic extension determines CTV and such extension can not be seen on CT scan this volume is usually based on the probability of microscopic disease being present and the importance of obtaining local control of such microscopic spread. The probability of microscopic disease being present is often estimated from knowledge of the pattern of spread of the tumor, histology, primary size, and other clinical parameters. The decision to include microscopic disease is not related to a single probability of involvement across all sites, however, the clinician must have a value in mind for each primary site. Subsequently, the CTV may need to be modified on the basis of surrounding normal structure tolerances.

The probabilistic nature of the PTV should be immediately apparent since it is based on organ motion and set up uncertainty. A simplistic determination of the margin for PTV around CTV in any one dimension is available by summing the variances of the errors randomly introduced by motion and set up variation in quadrature. The margin determined in this manner will include the CTV within the field 95% of the time. For example, assume the standard deviation of daily set up error for a tumor located in the thorax is 5 mm and the standard deviation of organ motion (including the CTV) due to respiration is 4 mm. Then the standard deviation of these random movements is $(5^2 + 4^2)^{1/2}$ or 6.4 mm and the CTV will be included 95% of the time if the PTV is drawn 1.3 cm beyond the CTV. This is admittedly a simplistic solution since it is based on unidimensional variations and these variations are often not well described. Nonetheless, this method provides a useful starting point for PTV determination. Recent data on the magnitude of set up variability and organ motion are referenced in Appendix I.

Prescription Doses for Target Volumes — The intended total dose and fractionation scheme for each CTV should be outlined. Additional information often helpful to the planner especially in non-routine cases includes the maximum and minimum CTV doses which would be acceptable. Whether or not it will be physically feasible to actually deliver the intended doses will be determined later in the planning process.

Anatomical Description and Dose Prescription for Dose Limiting Normal Structures — The dose limiting normal structures that will potentially be irradiated as part of the treatment should be specified along with their dose-volume relationship if known. Often the clinical data supporting the tolerance doses for all or partial organ irradiation are not well defined. The reader is referred to a paper by Emami et al.[7] which addresses current knowledge regarding late effect tolerance doses of a variety of normal tissues. Additionally, when chemotherapy or surgery are part of a combined treatment approach, the dose limiting normal structures may have an altered response to radiation which the clinician must keep in mind. Potentially, CT-assisted planning may produce extremely useful normal tissue tolerance information over time if dose volume data are accumulated and reported in a systematic way.

Suggested Beam Arrangements — At our institution, physicians are asked to communicate preferences for particular beam arrangements. Often, there are mitigating factors in a patient's medical history which are unknown to the planner but weigh heavily in the physician's choice of a particular beam arrangement. For example, a lung cancer patient may have both poor cardiac and pulmonary function which may greatly influence the design of an optimum treatment plan. Although these factors should have been communicated as part of the prescription for dose limiting normal structures, often the constraints for an individual patient's plan are naturally communicated when the physician and planner discuss possible beam arrangements.

IMMOBILIZATION

The first technical step in the treatment planning process is immobilization. The optimum treatment position and type of device will depend on several factors:

Patient's Medical Condition
As much as possible, the patient's comfort and their ability to maintain the treatment position must be considered.

Location of Disease and Probable Treatment Technique

The location of the patient's disease and the probable beam arrangements are primary considerations when choosing a position and immobilization method. Possible collisions with the treatment machine or CT scanner as well as undesirable radiation interactions with the immobilization device and the treatment couch must be avoided. If non-coplanar or unusual beam arrangements are being considered a simulation of the actual fields on the treatment machine may be necessary to ensure feasibility. Collision prediction by virtual simulators is an area of active development[8] and will be a helpful planning aid in the future.

Physical Compatibility with the CT Scanner

The patient must fit through the scanning aperture in the treatment position and not introduce unacceptable artifacts into the image.

The Fox Chase Cancer Center CT simulation planning worksheet used to record patient positioning and CT simulation information is shown in Appendix II.

LOCALIZATION

Localization is the crucial treatment planning step where the patient data set is obtained and all relevant aspects of the patient's anatomy including GTVs, CTVs, PTVs and dose limiting normal structures are defined geometrically with respect to a reproducible patient-defined coordinate system. Many of the details of localization using a CT simulator are discussed in Chapter 3.

During conventional simulation, the processes of localization and field design are combined. The physician designs and simultaneously images the actual treatment fields and, in doing so, implicitly defines the CTVs. Standard fields with apertures based on anatomical landmarks are routinely used for many sites. For cases requiring graphical isodose planning, target volumes are drawn on orthogonal films and transferred to manually drawn contours. The planner's task is to determine optimal wedges, weights, and, occasionally, beam angles, but apertures are usually still designed directly on radiographs of the treatment fields.

With the advent of CT assisted treatment planning, major transformations in localization and field design occurred. Localization has become distinct from field design which is now done using the virtual simulator and the treatment planning system. All field parameters, including the aperture, are designed to produce an optimum dose distribution — just the opposite of conventional simulation. Although this may seem a subtle difference, the implications for treatment planning are enormous as any department which has instituted CT planning has discovered. Physicians must learn to accurately delineate the patient's gross and presumed disease on the treatment planning CT. This requires them to not only improve their diagnostic skills but also to rethink the rationale of traditional radiation portals. Close cooperation with diagnostic radiologists and others who may be familiar with the tumor location (e.g., the surgeon) is required. At present, variations in GTV and CTV localization between physicians are relatively large due to the subjective interpretation of the CT images and the probabilistic nature of these structures. In the future, computer aided image correlation and other "intelligent" contouring aides may reduce this variation. Treatment planners must also obtain new skills such as three dimensional visualization and dose distribution optimization.

CT assisted planning is labor intensive for both physicians and treatment planners. If a CT simulator is not available, an initial simulation must be performed to define and mark the patient coordinate system and obtain orthogonal films for correlation with the CT data set. CT scans are then obtained in the treatment position and the physician localizes the disease and normal structures. After the graphical isodose planning is complete, the patient must return for another conventional simulation of the final treatment fields. Registration of the CT data set with the simulator images is a crucial step necessary to ensure an accurate transfer of the CT-based treatment plan to the treatment machine.

CT simulation offers several advantages over routine CT assisted planning which improve accuracy and efficiency. CT simulators have their own patient alignment and marking systems thus eliminating the need for the initial conventional simulation. Patients are scanned in the treatment position, the physician localizes the CTVs and subsequently a coordinate system centered on the CTV is defined and marked on the patient. Since the patient is marked after the CTVs have been defined, the final treatment isocenter rarely differs significantly from the coordinate system origin and, therefore, shifts from the origin or remarking of the patient are usually not necessary.

FIELD DESIGN/GRAPHICAL PLANNING

Field design is the treatment planning step during which all field parameters are determined, isodose distributions are produced when required, and all other treatment-related calculations are performed. Three major levels of field design complexity exist.

Simple treatment fields are those which can be designed based solely on anatomical landmarks unaided by CT-based target volume outlines. Combining the localization and field design steps into one process as was described above for conventional simulation is still an extremely efficient and desirable procedure for simple treatment fields. Efficient and cost effective use of a CT simulator demands that the capabilities exist to design simple fields such as whole brain portals in a manner analogous to conventional simulation, i.e., using DRR radiographic information only, unaided by CT localized contours. The CT simulator at our institution produces high resolution DRRs (512×512) which are sufficient for designing fields in this manner for most sites. The only region where spatial resolution is not currently sufficient using our standard scanning protocols is in the region of the vertebral bodies.

The next level of complexity in field design consists of standard treatment portals which benefit from CT localization of target volumes and normal structures but require little graphical isodose planning. Examples include anterior and posterior fields treating the mediastinum and primary disease in lung cancer or right and left lateral fields treating both nodal volumes and primary disease for head and neck sites. Using the virtual simulator, the fields are defined using the projection of the PTV in the region of the primary disease and the DRR radiographic information in the nodal regions. This procedure can usually be completed quickly, allowing the determination and marking of the final treatment field isocenter while the patient remains on the CT simulator couch. At this level of complexity, the CT simulator is used both as a conventional simulator and also as a CT localization tool. Figure 3 in Chapter 3 shows an anterior field designed for the treatment of a primary lung tumor. A combination of the primary PTV outline and bony anatomical landmarks are used to define a field aperture which includes the primary disease, the mediastinum and bilateral supraclavicular regions.

The most complex level of field design involves cases where all field parameters will be determined as part of the graphical isodose planning. Since the beam arrangement is not known a priori, fields are not designed while the patient waits on the scanner. Instead, GTVs,

CTVs, and PTVs are defined by the physician and reference points are defined in the center of the appropriate structures and marked on the patient. The patient is then dismissed. The advantages of CT simulation are most apparent at the complex level of field design. The CT virtual simulator is used to define an initial set of beams which geometrically maximize target coverage and minimize normal structure inclusion. Real time DRR display and modification of all treatment machine parameters make the virtual simulation analogous to conventional simulation with the advantage of superimposed CT-outlined structures. Beam direction, position, and aperture can all be defined and transferred along with the CT images, structure outlines and coordinate system definitions to the treatment planning system (TPS). This transfer can be accomplished in a number of ways, the most convenient being from the CT simulator to the TPS over a local area network (LAN). Graphical planning can then continue with the definition of additional beam parameters such as weighting and wedges and the calculation of isodose distributions. Further refinements to the beam are usually necessary before the finalization of the plan. At the present time, this is most efficiently done if there is a virtual simulator resident on the treatment planning system. However, in the near future, the bidirectional transfer of information between planning systems and CT simulators will become transparent to the user making virtual simulators on either device equally efficacious.

Once the treatment plan has been finalized, the CT virtual simulator is used to generate DRRs of the treatment fields including the apertures for comparison with the treatment machine portal images. An example of a conformally shaped field to treat the prostate and seminal vesicles is shown in Figure 4 of Chapter 3. The ability to produce a DRR for any arbitrary treatment field using the original CT localization data set is a great advantage of CT virtual simulation. Without the DRRs, the patient must return for a conventional simulation of the actual treatment fields; this incurs additional cost and introduces the possibility of misregistration between the CT data set and the new simulation.

VERIFICATION

The final step in the treatment planning process is verification of the treatment fields at the treatment machine. The DRRs of the actual treatment fields are used as the gold standard for comparison with treatment portal images. Annotation, documentation, and demographic information on the DRR, which is similar to that on simulation films, is essential. In most departments, hard copies of the DRRs are still the standard for comparison with portal films. At Fox Chase Cancer Center, all images including DRRs and portal images from electronic portal imaging devices are transferred over the LAN to a departmental wide PACS system. The DRRs and electronic portal images can then be simultaneously reviewed and approved by the physician using the computer at a remote station.

SUMMARY

The advantages of CT simulation for localization and field design are just beginning to be realized. The most obvious advantage is an increase in treatment planning efficiency. Standard CT-assisted planning often requires three localization procedures for each patient whereas CT simulation requires only one. Other advantages of CT simulation include the direct transfer of information from the CT simulator to other information systems within the radiation oncology department. CT images, coordinate system definitions, contour information, and field parameters can be transferred to and from the treatment planning system. Field apertures can be transferred directly to computerized block-cutting devices for the production of cerrobend blocks. In the near future, field parameters will be directly sent to record and verify systems. Picture archiving in radiation oncology is also in its early stages but will be advanced by the use of CT simulators.

At the present time, we feel there is a place for virtual simulators on both the treatment planning system and the CT simulator. Virtual simulation is most efficient on the CT simulator for the first two levels of field design complexity discussed above. For cases where the beam arrangement is determined as part of the graphical planning process, virtual simulation on the treatment planning system is extremely desirable. In the very near future, however, information will be transferred bidirectionally between the planning system and the CT simulator possibly obviating the need for virtual simulators on both. Bidirectional transfer of image and beam information between the CT simulator and various other information systems within the department holds great potential for improving the efficiency and accuracy of the treatment planning process.

REFERENCES

1. Leibel SA, Fuks Z, Zelefsky MJ, Whitmore WF Jr. The effects of local and regional treatment on the metastatic outcome in prostatic carcinoma with pelvic lymph node involvement. Int J Radiat Oncol Biol Phys 1994; 28:7-16.

2. Sandler HM, Perez-Tamayo C, TenHaken RK, Lichter AS. Dose escalation for stage C(T3) prostate cancer: minimal rectal toxicity observed using conformal therapy. Radiother Oncol 1992;23:53-54.

3. Leibel SA, Heimann R, Kutcher GJ, Zelefsky MJ, Burman CM, Melian E, Orazem JP, Mohan R, LoSasso TJ, Lo YC, Wiseberg JA, Chapman DS, Ling CC, Fuks Z. Three-dimensional conformal radiation therapy in locally advanced carcinoma of the prostate: preliminary results of a phase I dose-escalation study. Int J Radiat Oncol Biol Phys 1994;28:55-65.

4. Hanks GE, Lee WR, Schultheiss TE. Clinical and biochemical evidence of control of prostate cancer at five years after external beam radiation. J Urology 1995;154:456-459.

5. Boyer AL, Antonuk L, Fenster A, Van Herk M, Meertens H, Munro P, Reinstein LE, Wong J. A review of electronic portal imaging devices. Med Phys 1991;19:1-16.

6. ICRU Report 50. Prescribing, Recording, and Reporting Photon Beam Therapy. Bethesda, Maryland: International Commission on Radiation Units and Measurements, 1993.

7. Emami B, Lyman J, Brown A, Coia L, Goitein M, Munzenrider JE, Shank B, Solin LJ, Wesson M. Tolerance of normal tissue to therapeutic irradiation. Int J Radiat Oncol Biol Phys 1991; 21:109-22. Review.

8. Humm JL, Mohan R, Caley R, Fleischman R, Kutcher GJ. Methods to detect machine/patient collisions in computer controlled multi-segment therapy. Med Phys 1994;21:940. Abstract.

— *Part II* —

Clinical Applications *of* CT Simulation

— *Chapter 7* —

ESOPHAGEAL CANCER

Lawrence R. Coia, MD

INTRODUCTION

Esophageal cancer is diagnosed in 12,000 people in the United States annually and results in over 10,000 deaths.

Concurrent chemotherapy and radiation treatment has produced superior local control and survival compared with radiation alone.[1] Nonetheless, local failure with chemoradiation occurs in 35% to 45% of patients.[1,2] Therefore, improvements in local control are needed. Furthermore, chemoradiation as delivered in the Intergroup trial reported by Herscovic et al. results in a relatively high rate (33%) of acute Grade 3 or 4 upper GI toxicity. To some extent, this high toxicity rate may be related to the large fields used. For example, for patients treatment with chemoradiation, the initial volume (treated to 30 Gy) extended from supraclavicular nodes to the GE junction (except for lower third lesions where the supraclav was excluded), and the boost volume (treated to 50 Gy) included 5 cm above and below the gross tumor. Careful delineation of the CTV and improved treatment planning via CT simulation may make possible an increase in local tumor control through delivery of higher radiation doses and decrease in treatment toxicity through use of smaller, accurately placed fields.

In this chapter, we will describe the use of CT simulation for primary management of patients with esophageal cancer. The use of CT simulation in the postoperative setting is not discussed since chemoradiation is used almost exclusively as the primary treatment or in the preoperative setting for esophageal cancer at the Fox Chase Cancer Center.

ANATOMY

Anatomically, the esophagus represents a difficult treatment planning problem for the radiation oncologist. It is a lengthy organ extending from the cricopharyngeal muscle in the neck to the GE junction (see Figure 1). The length of the esophagus is roughly 25 to 30 cm. The cervical esophagus starts at the cricopharyngeal muscle and extends to the level of the thoracic inlet. The thoracic esophagus is divided into three sections, the upper thoracic esophagus extends from the thoracic inlet to the carina, the mid thoracic esophagus starts at the carina and extends halfway to the GE junction and the lower thoracic esophagus extends from that point to the GE junction. The cervical esophagus is the shortest segment measuring approximately 5 to 7 cm. The location of the lesion in the esophagus is usually described in relationship to its distance from the incisors measured at the time of endoscopy. As measured from the incisors, cervical esophageal lesions are generally located at 13 to 18 cm, upper thoracic lesions at 18 to 24 cm, mid thoracic lesions at 24 to 32 cm, and lower thoracic lesions at 32 to 40 cm. The course of the esophagus is vertically downward, but there are mediolateral and dorsal-ventral curvatures. It initially deviates to the left of midline in the cervical esophagus and remains to the left through the root of the neck gradually reaching the midline at the fifth thoracic vertebrae. Caudally, it shifts to the left of midline and is more ventrally located at the esophageal hiatus in the diaphragm. The dorsoventral flexures of the esophagus correspond to the curves of the vertebral column.

The esophagus is surrounded by a number of dose-limiting anatomical structures including lungs, heart, and spinal cord. Since distal lesions often extend into the stomach and celiac nodal regions are at high risk for tumor involvement, dose-limiting upper abdominal structures such as liver, kidneys, and stomach must also be taken into consideration. Important relationships of the esophagus to surrounding anatomic structures are diagramed in Figure 1.

Intraluminal contrast aids in visualization of the esophagus on CT scan, however, occasionally there is total esophageal obstruction which limits passage of contrast. Therefore, it is useful to have a good understanding of the anatomic relationship shown in Figure 1 in order to better estimate esophageal location. The cervical esophagus is bounded ventrally by the trachea with the recurrent nerves ascending in the groove between the trachea and the esophagus, dorsally by the vertebral column and prevertebral musculature and laterally by the thyroid gland and carotid arteries. The thoracic esophagus is bounded ventrally by the trachea in its upper portion, by the left bronchus in its mid portion, by the pericardium, left vagus and

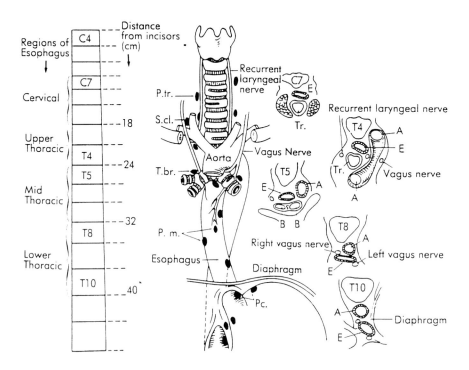

Figure 1. Diagram of important relations of the esophagus. These structures are invaded early and frequently should be included in the volume of irradiation. P.tr. indicates paratracheal; S.cl., supraclavicular; T.br., tracheobronchial; PM, posteromedial; P.c., paracardial.[10]

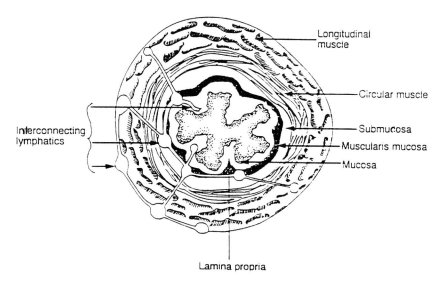

Figure 2. An exaggerated cross-section of the esophagus, illustrating its five layers and the interconnected submucosal and muscular lymphatics that course the length of the esophagus.[10]

diaphragm in its lower portion. Dorsally it is bounded by the vertebral column and longus colli muscle with the intercostal arteries, thoracic duct, right vagus and aorta at its caudal end. To the left of the thoracic esophagus is the aortic arch, left subclavian artery and thoracic duct in its uppermost aspect, and the pleura and descending aorta in the mid and lower thoracic esophagus. To the right of the esophagus are the azygous vein and pleura along with right vagus nerve and thoracic duct.

The normal esophagus is comprised of mucosa and muscular wall without a serosa. A cross-sectional diagram of the esophagus is shown in Figure 2. An understanding of this aspect of the esophageal anatomy is essential since the present staging system is based on depth of wall penetration. There is a rich interconnecting lymphatic network in the mucosal, submucosal and muscular wall region which allow both longitudinal and transmural spread of tumor.

PATTERNS OF SPREAD

Esophageal cancer kills by both local progression and distant metastases. In the United States less than 20% of patients present with disease confined to the esophagus. The majority of patients have regional lymph node involvement and/or distant metastases. The regional lymph nodes at highest risk for involvement are shown in Figure 3. Most of the data available regarding regional nodal spread from esophageal cancer comes from esophagectomy series which have included patients who have had squamous cell cancer rather than adenocarcinoma of the esophagus.[3] Adenocarcinoma of the esophagus also has a high rate of local regional failure, however, an accurate mapping of lymph nodes at risk as has been done for squamous cell cancer of the esophagus is not available for adenocarcinoma. The probability of lymph node involvement as a function of the site in the esophagus of the primary tumor is tabulated in Table 1. Mediastinal, celiac and gastric nodes are at risk for lesions at or below the carina while supraclavicular and mediastinal nodes are at risk for tumors arising in the upper and

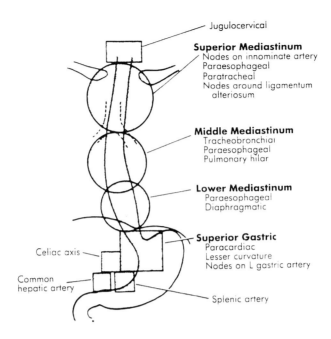

Figure 3. Lymph node groups that drain the esophagus. (From Thompson WM: Int J Radiat Oncol Biol Phys 9:1533, 1983. After Akiyama H et al: Ann Surg 194:438-446, 1981.)

mid-esophagus. The location of the celiac nodes relative to vertebral bodies have been described by Kao et al.[4] and are shown diagrammatically in Figure 4. Note that in general these nodes are above the pedicles of L1.

DIAGNOSTIC STUDIES

All patients should undergo barium swallow, endoscopy with biopsy, and a CT scan of the thorax and upper abdomen to localize the lesion. Endoscopic ultrasound is complementary to CT scan, having an 85% accuracy in determining depth of wall penetration (T stage) and 75%

Table 1. Probability of Lymph Node Involvement as a Function of the Site in the Esophagus of the Primary Tumor

Location of Primary	Supra-clavicular	Nodes Groups at Risk Mediastinum			Gastric	Celiac
		Superior	Mid	Lower		
		Percentage of Involvement				
Thoracic						
Upper	40	30	28	29	32	0
Middle	25	11	21	18	33	4
Lower	10	10	14	27	61	21

Celiac Axis

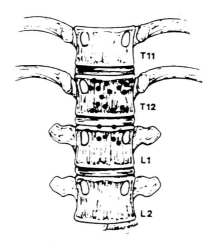

Figure 4. Graphical representation of the specific locations of origin of the celiac axis as determined from patient angiograms.[4]

accuracy in determining periesophageal lymph node involvement (N stage), while CT has a 90% accuracy in detecting pulmonary and hepatic metastasis (M stage).[5] The use of minimally invasive surgical procedures such as thoracoscopy and laparoscopy may provide important information in determining the areas at risk for esophageal tumor. However, at present these studies remain investigational.[6] A comparison of staging systems for the primary esophageal cancer is shown in Table 2.

As previously mentioned, the present staging system is based on depth of wall penetration and nodal involvement. In the present staging system, celiac nodal involvement represents distant metastases for tumors of the thoracic esophagus. The prior staging system was based on tumor length, obstruction, degree of circumferential involvement and presence or absence of

Table 2. A Comparison of Staging Systems for the Primary Esophageal Cancer

T Stage	1988 AJC	1983 AJC	CT System
T1	Into lamino propria or submucosa	<5 cm, no obstruction not circumferential	Intraluminal mass
T2	Into muscularis propria	>5 cm, obstruction or circumferential	Mass and wall thickening >5 mm
T3	Into adventitia	Clinical extra-esophageal spread	Into adjacent structures (distant mets)
T4	Into adjacent structures		

extraesophageal spread. It is likely that a combination of these staging systems would provide better prognostic information for patients managed with chemoradiation as a primary treatment modality.

IMMOBILIZATION

Patient Positioning

For patients with tumors of the thoracic esophagus it is generally recommended that the patient be in the prone position throughout. This is sometimes difficult if patients have had a prior gastrostomy or jejunostomy tube placed. Patients should be immobilized in an Aquaplast cast, and, if a tube has been placed, the cast can be formed to accommodate it. Patients with cervical or cervical and upper thoracic esophagus cancers can be in the supine position. A bite block may be useful for patients with cervical lesions. Positioning of the arms to the side of the patient is generally most comfortable, however, if lateral fields must be used then positioning of the arms above the head may be required.

LOCALIZATION

Definition of Gross Tumor Volume

Gross tumor volume (GTV) represents tumor in the esophagus or lymph nodes seen on diagnostic studies. Esophageal wall thickening on CT scan greater than 5 mm is not normal and may indicate tumor involvement. Loss of the normal fat plane seen between the esophageal tumor and surrounding structures may also indicate extraesophageal spread; however, patients with esophageal cancer are often in poor nutritional status and the fat plane may be difficult to discern. The use of contrast material at the time of CT simulation will help differentiate the esophagus from surrounding structures such as aorta and trachea. Although nodes greater than one-half cm in the periesophageal region on CT scan are likely to be involved with tumor, CT scan is not as accurate (only 56% accuracy) as endoscopic ultrasound in determining gross or microscopic evidence of periesophageal lymph node involvement (75% accuracy). We use a 1 cm proximal and distal margin around the gross tumor to identify the GTV due to inaccuracies in CT scan in determining the gross tumor length.

Definition of Clinical Target Volume

The clinical target volume (CTV1) represents the gross target volume along with microscopic extension of disease. The microscopic extension of disease may be predicted from knowledge of the spread of tumor as determined by data from surgical series. For example, the incidence of microscopic involvement of the proximal and distal margins of resection of the esophagus is a function of the distance of the axial resection margin from the primary tumor. In the study of Tam et al.[7] a 4 to 6 cm margin was associated with a 15% rate of microscopic involvement compared to only 8% microscopically positive margins with a 6 to 8 cm resection margin. We therefore recommend that the initial clinical target volume include the GTV plus a minimum of 6 cm above and below the primary tumor along the axis of the esophagus.

The radial margin is rarely mentioned in pathologic reports; however, since over 60% of patients who undergo "curative" esophageal resections are noted to have tumor penetrating the entire esophageal wall, we suggest a minimum of 1.0 cm radially beyond esophageal wall or tumor be used for the CTV in order to encompass radial microscopic spread. The high potential for regional lymph node involvement with microscopic disease must also be considered when determining the CTV1. In general, when there is greater than 15% incidence of microscopic nodal likelihood, we include those nodal regions in the CTV1. The following policy regarding lymph node inclusion is used (see Table 3): cervical esophagus — cervical and mediastinal nodes; upper thoracic esophagus — supraclavicular and mediastinal nodes; mid-thoracic esophagus — mediastinal and celiac nodes ± supraclavicular nodes; lower thoracic esophagus — mediastinal and celiac nodes. Clinical target volume 2 (CTV2) corresponds to the GTV.

Definition of Planning Target Volume

The planning target volumes represent the clinical target volume plus a margin for uncertainty based on motion of the esophagus and set up uncertainty. The near midline location of the esophagus in the thorax makes it relatively stable, so that its position is not nearly as affected by respiration as more laterally-based tumors. We estimate that motion of the esophagus due to respiration and esophageal motility averages less than 2 mm in superior, inferior or right-left (lateral) direction, and 4 mm in anterior or posterior direction when the patient is set up in the prone position. Similar assumptions can be made for the regional thoracic nodes. Set up uncertainty should average 5 mm (or less) in this casted position. Adding motion and set up errors in quadrature and multiplying by two (e.g., $2\sqrt{5^2 + 4^2}$) gives the linear distance from CTV to PTV which will encompass the CTV 95% of the time. Therefore, the planning target volume (PTV1 or PTV2) encompasses the clinical target volume (CTV1 or CTV2) if a

Table 3. Definition of Target Volumes by Site of Primary

GTV = Gross disease seen on CT scan plus 1 cm proximal and distal margin
CTV1 = GTV + 6 cm proximal and distal margin
          + 1 cm radial margin
          + nodes (see chart below)
CTV2 = GTV
PTV1 = CTV1 + 1.3 cm
PTV2 = CTV2 + 1.3 cm

Nodal Regions Included by Site of Primary in the CTV1

Site of Primary	Nodal Regions
Cervical	Cervical, mediastinal (superior and middle)
Upper thoracic	Supraclavicular, mediastinal (superior and middle)
Middle thoracic	Mediastinal (superior, middle and lower), celiac
Lower thoracic	Mediastinal (middle, lower), celiac
GE junction	Mediastinal (middle, lower), celiac

margin in superior, inferior and lateral extent and a 1.3 cm margin anteriorly and posteriorly were used. At present, we have chosen to use a 1.3 cm margin beyond CTV in all direction to define PTV for cancer of the esophagus. The definition of target volume by site of primary are shown in Table 3.

Definition of Critical Structures

Critical structures when treating esophageal cancer may include the spinal cord, stomach, heart, lungs, and kidneys. The tolerance doses of these organs are also outlined in Table 4. For esophageal cancer, we limit critical structures to the following doses: spinal cord 45 Gy, stomach 60 Gy (to less than one-third), heart 45 Gy (to less than one-half), lungs 2000 cGy (to less than one-third). The tolerance of these structures may be affected by the administration of concurrent chemotherapy and the type of chemotherapy given. Placement of the patient in the prone position will displace the PTV2 away from the spinal cord,[8] however, prone positioning may result in slightly more heart in the PTV. Since the PTV2 moves 1 to 2 cm away from the spinal cord, and the increase in heart volume in the field is <20%, we prefer prone positioning for all tumors at or below the carina.

CT Scan Localization Optimization & Virtual Simulation

For cervical esophageal lesions the superior scanning limit is at the tip of the mastoid bone and inferior scanning limit at the carina. For lesions of the upper thoracic esophagus the superior limit is the thyroid notch and the inferior limit is the diaphragm. For mid-thoracic lesions the superior limit is the thyroid notch and inferior limit is L1-L2. For lower thoracic lesions the superior limit is the aortic arch and inferior limit is L1-L2. Occasionally these scanning limits may be modified as one must take into account specifics of the situation such as bulky adenopathy, extraesophageal extension, or lengthy tumors. The slice spacing should be 3 mm and slice thickness 5 mm. 50 cc of oral CT barium (1.2% barium sulfate) mixed with 15 cc of esophatrast are given immediately prior to scanning. This mixture does not significantly degrade image quality and can be quite useful in localizing the esophagus and esophageal tumor.

Treatment localization and CT simulation procedures are performed by the therapist (technologist) using the general guidelines outlined in Chapter 3. For esophagus specifically, the isocenter is defined in the treatment localization mode using the CTV1 outlined by the physician. The therapist outlines critical structures including spinal cord and often other normal structures such as carina for reference structure (see Chapter 3). Beam arrangement optimization (part of the virtual simulation process) is then carried out by the dosimetrists and/or physician. DRRs of the planned beam fields with target, critical structures, and selected normal structures outlined are obtained and conformal beam blocks are drawn using the treatment planning guidelines mentioned earlier.

TREATMENT PLANNING

Dose Prescription for Clinical Target Volume and Critical Structure

The following guidelines for radiation dose prescription were formulated on the basis of delivery of concurrent chemotherapy (infusional 5 Flououracil ± Mitomycin or ± Cisplatinum)

Table 4. Radiation Tolerance of Normal Tissues Encountered in Radiation of Esophageal and Gastric Cancer

Organ	Injury	TD 5/5 (Gy)	TD 50/5 (Gy)	Segment
Class I Organs				
Liver	Hepatitis	25	40	Whole
Stomach	Perforation, ulcer	45	55	100 cm^2
Intestine	Perforation, ulcer	45	55	400 cm^2
	Perforation, ulcer	50	65	100 cm^2
Spinal Cord	Infarction, necrosis	45	55	10 cm
Heart	Pericarditis	45	55	60%
Lung	Pneumonitis	15	25	Whole
		30	35	100 cm^2
Kidney	Nephrosclerosis	20	25	Whole
Class II Organs				
Esophagus	Esophagitis, ulceration	60	65	
Adrenals	Hypoadrenalism	>60	...	
Class III Organs				
Muscle (adult)	Fibrosis	60	80	
Large arteries and veins	Sclerosis	>80	>100	

and radiation therapy and are presented in Table 5. The dose to the initial planning target volume (PTV1) is 30 to 40 Gy. This dose range when used in combination with concurrent chemotherapy should be sufficient to provide microscopic disease control. The initial PTV1 is often treated by initial anterior and posterior fields to 30 Gy followed by an anterior and two posterior obliques for up to an additional 10 Gy. The PTV2 is treated to 60 Gy. Generally this second volume is treated via an anterior field in combination with two posterior obliques. Establishment of the isocenter for PTV1 and PTV2 can be made on the basis of the initial CTVs outlined.

Dose volume histogram analysis of two frequently used techniques indicates potentially less pulmonary toxicity with the technique described here compared to the use of two-posterior obliques and an anterior field (three-field) throughout treatment since a larger volume of lung receives a higher dose of radiation with the latter technique.[8] However, there is a slightly smaller volume of heart receiving a higher dose with the use of three-field technique throughout.

Table 5.

Dose Prescription*	
PTV1	40 Gy
PTV2	60 Gy

*Generally encompassed by 95% isodose or higher.

TREATMENT VERIFICATION

Digitally reconstructed radiographs (DRRs) for patients with esophageal cancer can be used to verify field arrangements by using bony landmarks. However, currently the DRRs often provide insufficient information regarding soft tissues of importance in the thorax and upper abdomen. Therefore, at the present time, in addition to performing a CT simulation, most patients also undergo a conventional simulation with barium contrast. The target volume information generated by the CT simulation is easily transferred to the conventional simulation fields, and the position of the esophagus can be verified. In addition, the critical structures such as spinal cord, kidneys, or liver can be delineated.

CASE ILLUSTRATIONS

Patient #1

Lower cervical and upper thoracic esophageal cancer, Stage T3N1M0 (1988, AJC). The treatment of the lower cervical and upper thoracic esophagus represents a formidable problem for radiation oncologists since the target volume includes lymph nodes both in the neck and mediastinum, the critical structures such as spinal cord, heart, and lungs are often quite close to the target volume, and there are rapidly changing external contours from the neck to the thorax. The gross target volume for this patient with a 5 cm tumor in the low cervical upper thoracic region is represented in Figure 5. Three millimeter slices were used in scanning the patient. Endoscopic ultrasound demonstrated tumor extending through the esophageal wall with involvement of two periesophageal lymph nodes. A cross-sectional view of the GTV, CTV and PTV at central axis is shown in Figure 5a. GTV is blue. The CTV1 representing GTV plus microscopic disease is illustrated in green. In this case, CTV1 was drawn by using a 1 cm radial margin beyond gross, primary, and suspected lymph node involvement, an approximate 6 cm superior and inferior margin on the esophageal tumor, plus inclusion of supraclavicular and mediastinal nodal regions. The PTV1 in purple was drawn by adding an additional 1.3 cm margin beyond the CTV1 for motion and set up error. The PTV2 in orange is approximately 1.3 cm around grossly involved sites. However, in some areas the border of the PTV2 is less than 1 cm from gross disease based on anatomic structures which limit tumor invasion and restrict motion (e.g., vertebral body). An anterior DRR is also illustrated (Figure 5b). The DRRs provide an adequate visualization of bony structures, however, a conventional simulation was necessary in order to better define soft tissues (including the esophagus) and to verify that the custom designed blocks were not blocking tumor and were blocking appropriate normal tissues.

Patient #2

Adenocarcinoma of the distal esophagus, stage T2NXM0 (1983, AJC). This patient had a 10 cm lesion of the mid- and distal esophagus with no extra-esophageal extension. In Figure 6a, which shows a slice at the central axis, the spinal cord is delineated in green, carina in red, GTV in purple, CTV1 in blue PTV1 in yellow, PTV2 in orange. One cm scanning slices were used. There is preservation of the plane between the esophagus and aorta. The CTV1 is 6 cm beyond the gross target volume superiorly and inferiorly and a minimum of 1 cm radially, and additionally the mediastinal and gastric/celiac nodal regions were included. The PTV1 in yel-

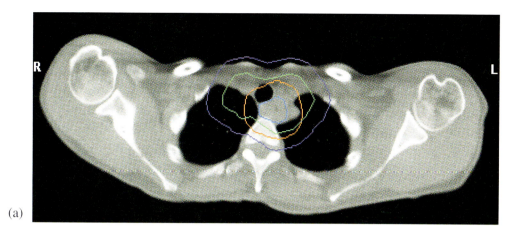

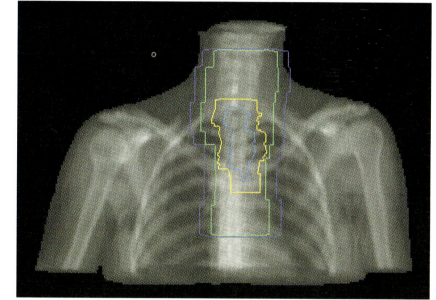

Figure 5. Patient #1 with squamous cell carcinoma of lower cervical and upper thoracic esophagus. GTV: blue; CTV1: green; PTV1: purple; and PTV2: orange. (a) Axial CT scan showing target volumes. (b) Anterior DRR showing initial and cone down volumes.

low was an additional 1.3 cm beyond the CTV1. The PTV2 (cone-down volume) represents GTV with a 1.3 cm margin. An anterior view DRR with CTV, PTV, GTV, carina and spinal cord segment is shown in Figure 6b as well as an anterior DRR with GTV and PTV2, spinal cord and carina in Figure 6c. Although esophatrast was quite useful to outline the esophagus in each slice, the esophatrast is not well seen on the DRR. A conventional simulation was also performed in order to verify block position.

CONCLUSION

Optimization of the therapeutic ratio in esophageal cancer can be achieved by an improvement in tumor control and a decrease in treatment toxicity.[9] CT simulation and treatment planning offer significant advantages in target volume definition over conventional simulation.

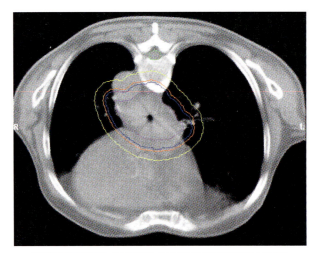
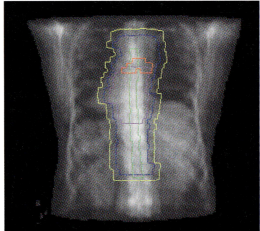
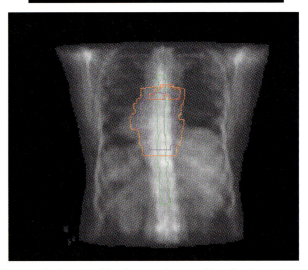

Figure 6. Patient #2 with adenocarcinoma of the distal esophagus. GTV: purple; CTV1: blue; PTV1: yellow; PTV2: orange; spinal cord: green; and carina: red. (a) Axial CT showing target volumes. (b) Anterior DRR showing initial and cone down volumes. (c) Anterior DRR showing cone down volumes.

This should allow for delivery of high radiation doses to the target volume and lower doses to surrounding critical structures. It is likely that 10% to 20% improvements in local tumor control through delivery of a higher radiation dose and a similar percentage decrease in toxicity through critical structure delineation are possible through optimized radiation treatment planning via CT simulation. At present, conventional simulation is often needed to verify block position, etc., as determined by CT simulation; however, with improvements in DRR and ability to delineate important critical structures, this may not be necessary.

REFERENCES

1. Herscovic A, Martz K, Al-Sarraf M, Leichman L, Brindle J, Vaitkevicius V, Cooper J, Byhardt R, Davis L, Emami B. Combined chemotherapy and radiotherapy compared with radiotherapy alone in patients with cancer of the esophagus. N Eng J Med 1992;326:1593-1598.

2. Coia LR, Engstrom P, Paul A, Stafford P, et al. Long-term results of infusional 5-FU, mitomycin-C and radiation as primary management of esophageal carcinoma. Int J Radiat Oncol Biol Phys 1991;20:29-6.

3. Akiyama H, Tsurumaru M, Kawamura T, Ono Y. Principles of surgical treatment for carcinoma of the esophagus: analysis of lymph node involvement. Ann Surg 1981;194:438-446.

4. Kao GD, Coia LR. Anatomy of the celiac axis and superior mesenteric artery and its significance in radiation therapy. Int J Radiat Oncol Biol Phys 1992;25:131-134.

5. Lightdale C. Staging of esophageal cancer I: endoscopic ultrasonography. Semin Oncol 1994;21:438-446.

6. Jaklitsch M, Harpole D, Healy E, Sugarbaker D. Current issues in the staging of esophageal cancer. Semin Radiat Oncol 1994;4:135-145.

7. Tam PC, Siu KF, Cheung HC, Ma L, Wong J. Local recurrences after subtotal esophagectomy for squamous cell carcinoma. Ann Surg 1987;205:189-194.

8. Corn B, Coia L, Chu J, Huang C, Hanks G. Significance of prone positioning in planning treatment for esophageal cancer. Int J Radiat Oncol Biol Phys 1991;21:1303-1310.

9. Gaspar L. Radiation therapy for esophageal cancer: improving the therapeutic ratio. Semin Radiat Oncol 1994;4:192-201.

10. Coia LR. The esophagus. In: Cox JD, ed. Moss's Radiation Oncology: Rationale, Technique, Results. St. Louis, Missouri: C.V. Mosby, 1994, chapter 16.

Chapter 8

HEAD & NECK CANCER

Douglas A. Fein, MD, Andrew Shaer, MD, and Kiaran P. McGee, MS

LARYNX

INTRODUCTION

The use of CT treatment planning in laryngeal cancer may improve the therapeutic ratio by maximizing the dose to the tumor while minimizing the dose to critical structures. The larynx has a steep dose response curve for both tumor control and normal tissue complication probability. Treatment planning is difficult due to the changing contour from the head through the neck and the thorax. In addition, dose inhomogeneities may arise due to air in the upper aerodigestive tract. Radiation therapy plays an important role in the treatment of laryngeal cancer and thus optimal treatment planning and the use of a CT simulator may prove to be an important advance for T2-T4 tumors of the larynx.

ANATOMY

Anatomically, the larynx can be divided into the supraglottic larynx, glottic larynx, and the subglottic larynx. The division between the glottic larynx and supraglottic larynx is the point where the epithelium of the true vocal cord turns upward to form the ventricle. The subglottic larynx begins 5.0 mm below the free margin of the true vocal cord. The supraglottic larynx can be subdivided into the epiglottis, false vocal cords, ventricles, aryepiglottic folds and arytenoids. The glottic larynx is comprised of the true vocal cords and the anterior commissure (Figure 1).

The larynx is supported by bone, cartilage, and ligaments. The outside of the larynx is comprised of the hyoid bone, thyroid cartilage and cricoid cartilage. The anterior aspect of the larynx is comprised of the epiglottis, arytenoids, corniculate and cuneiform cartilages (Figure 2). Important avenues for cancer spread include the pre-epiglottic space and paraglottic space. Anatomically, these fat spaces are defined by the hyoepiglottic ligament superiorly, the conus elasticus inferiorly, the thyroid ala, thyrohyoid membrane, and hyoid bone anteriolaterally, medially by the intrinsic muscles, and posteriorly by the anterior wall of the pyriform sinus.[1,2]

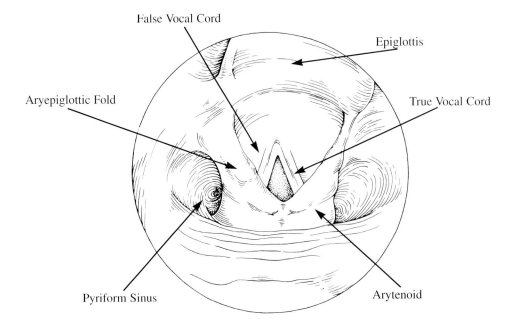

Figure 1. Artist's rendition of a normal larynx.

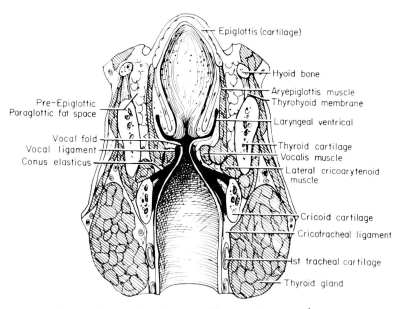

Figure 2. Coronal diagram showing normal laryngeal anatomy.[1]

PATTERNS OF SPREAD

Glottic Larynx
In general, lesions of the glottic larynx may spread to the:
a. anterior commissure
b. Broyle's ligament (anterior commissure tendon)
c. subglottic larynx
d. ventricle/false vocal cord

At the time of diagnosis, approximately two-thirds of glottic larynx tumors are confined to the true vocal cord. Of note, tumors involving the anterior commissure may extend anteriorly along Broyle's ligament to invade the thyroid cartilage.

Supraglottic Larynx
Potential avenues of spread for supraglottic structures depend primarily on the initial structure involved.

Epiglottis
1) vallecula
2) pre-epiglottic space
3) lateral pharyngeal wall
4) base of tongue
5) false vocal cords
6) aryepiglottic folds

Aryepiglottic Folds/Arytenoids
1) epiglottis
2) thyroid cartilage
3) cricoid cartilage
4) base of tongue
5) pharyngeal wall
6) pyriform sinus

False Vocal Cord/Ventricle
1) paraglottic space
2) medial wall of pyriform sinus
3) true vocal cord
4) infrahyoid epiglottis

Lymphatic Spread

Glottic Larynx
Due to the sparse lymphatic drainage of the glottic larynx, lymph node involvement is uncommon for early glottic cancer. The incidence of clinically or pathologically positive nodes for patients presenting with T1 lesions is 0%, 2% to 5% for individuals with T2 lesions, and 20 to 30% for patients presenting with T3 or T4 lesions.

The nodes at risk in patients presenting with primary glottic carcinomas are shown in Figure 3 and include:
   a. jugulodigastric nodes, particularly if there is supraglottic spread
   b. lateral paratracheal nodes
   c. midline pretracheal node (Delphian node)
   d. mid and low jugular nodes

Supraglottic Larynx

Primary lymphatic drainage of the supraglottic larynx structures is to the subdigastric nodes, as well as the upper and midjugular nodes. At the time of diagnosis, 55% of patients have positive nodes and 15% of individuals present with bilateral lymphadenopathy. Thirty-three percent of individuals who present with an N0 neck and are observed will develop positive nodes.

STAGING

Glottic Larynx
- T1: tumor limited to the true vocal cord(s) with normal mobility.
- T2: tumor extends to the supraglottic and/or subglottic and/or with impaired vocal cord mobility.
- T3: tumor limited to the larynx with vocal cord fixation.
- T4: tumor invades the thyroid cartilage and/or extralaryngeal extension.

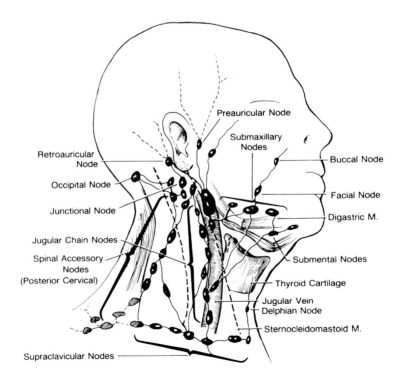

Figure 3. Lymph nodes of the head and neck.[1]

Supraglottic Larynx

T1: tumor is limited to one subsite.
T2: tumor invades more than one subsite.
T3: tumor limited to the larynx with vocal cord fixation and/or invades postcricoid area, medial wall of pyriform sinus, or pre-epiglottic space.
T4: tumor invades through the thyroid cartilage or extralaryngeal spread.

DIAGNOSTIC STUDIES

Staging studies should include: 1) Indirect laryngoscopy; 2) CT with contrast or MRI of the larynx and entire neck from the base of skull to several centimeters below the clavicle; 3) direct laryngoscopy with biopsy; 4) chest x-ray.

IMMOBILIZATION

Patient Positioning

The patient should be positioned with his/her head in a neutral position, i.e., hard palate should be parallel to the plane of the true vocal cord. This ensures that vocal cord tissue and larynx are intersected perpendicularly by the axial CT slice, reducing the possibility of oblique dissection of these organs.

All patients should be treated supine with their heads in a head holder that permits adequate extension. Immobilization is essential and this should be accomplished utilizing either an Aquaplast mask or a bite block technique. Advantages of Aquaplast immobilization include patient comfort and lack of need to allow for healing of the oral cavity if dental extractions were required prior to treatment. Advantages of the bite block include excellent immobilization and the ability to displace the tongue out of the field by use of the mouthpiece if necessary.

LOCALIZATION

Gross Tumor Volume / Clinical Target Volume

The gross tumor volume (GTV) is equated with findings on the CT scan as well as the physical examination. Findings on clinical evaluation are correlated with information derived from axial CT or MRI data. As a result, at the Fox Chase Cancer Center the clinical tumor volume (CTV) is defined at the time of CT simulation and the GTV is encompassed in this volume. There are a total of seven CTVs that are typically treated and are listed below.

CTV 1. Primary Tumor
CTV 2. Cervical nodes
CTV 3. Subdigastric nodes
CTV 4. Posterior cervical nodes
CTV 5. Supraclavicular nodes
CTV 6. Upper mediastinal nodes
CTV 7. Infraclavicular nodes

Planning Target Volume

The planning target volume (PTV) should include the CTV plus a margin for organ motion and set up uncertainty. Previous studies have reported set up uncertainties (one standard deviation) of 1.5 mm,[3] 2.1 mm,[4] and 4.0 mm[5] for head and neck treatments. For 95% coverage, a margin of two standard deviations around the CTV is used to define the PTV. Using these figures, margins can accordingly range from 3.0 mm to 8.0 mm. Four different PTVs are used to treat the larynx at the Fox Chase Cancer Center and are summarized below.

PTV1 = CTV1 + CTV2 + CTV3 + CTV4 + 3.0-8.0 mm margin
PTV2 = CTV5 + CTV6 + CTV7 + 3.0-8.0 mm margin
PTV3 = CTV1 + CTV2 + CTV3 + 3.0-8.0 mm margin
PTV4 = CTV4 + 3.0-8.0 mm margin

Definition of Critical Structures

Various normal structures are typically included within the irradiated volume as a result of selected beam geometry and proximity to the tumor. Tolerances of normal tissues irradiated can be defined by their TD 5/5 which gives the maximal dose that can be delivered to the organ with minimal (5%) risk of complications. Table 1 lists the various normal tissues at risk in the treatment of laryngeal cancer and their respective TD 5/5 values.[6]

CT Scan Localization/Optimization & Virtual Simulation

Scanning Borders

The CT scan volume should encompass the target volume and enough normal tissue to ensure adequate anatomical information for reconstructing digitally reconstructed radiographs (DRRs). The superior border should cover the external auditory meatus while the inferior border should extend to the bottom of the cricoid cartilage if the supraclavicular fossa is not treated. Otherwise, the inferior border should be the top of the aortic arch.

Table 1. Normal Tissues at Risk and their Respective TD 5/5 Values

Organ	TD 5/5 (Gy)	Complication
Parotid	32	Xerostomia
Soft tissue / bone	60	Bone exposure/ osteoradionecrosis
Larynx	70	Chondronecrosis
	45	Edema
Spinal cord	45	Myelitis
Skin	70	Fibrosis (<10 cm^2)
	60	Fibrosis (<30 cm^2)
	55	Fibrosis (<100 cm^2)
External auditory meatus	30	Acute serous otitis
	55	Chronic serous otitis
Brachial plexus	60	Plexopathy

Image Technique

Image with bone windows at the level of the thyroid cartilage.

Slice Protocol

CT slice thickness and separations should be selected to ensure adequate resolution of the target volume and normal structures in DRRs. Images should be obtained at 2 mm thicknesses with increments of 2 mm in order to provide optimal visualization and subsequent reconstruction in the form of DRRs of these structures.

Contrast

100 to 150 cc of iodinated intravenous contrast should be administered by bolus and continuous drip infusion immediately before and during the CT scan in order to adequately visualize the vessels and nodes.

Localization and Virtual Simulation

Treatment localization and CT simulation procedures are performed by the therapist using the guidelines outlined in Chapter 3. For the larynx, the treatment isocenter is defined as the center of CTV1 as outlined by the physician. Other structures such as CTV2- CTV7, and critical normal ones such as spinal cord are outlined by the physician and/or dosimetrist. Beam arrangement (virtual simulation) is then performed by the dosimetrist and physician. DRRs of the planned fields are regenerated and conformal blocks added.

TREATMENT PLANNING

Radiation should be administered in fraction sizes of 1.8 to 2.0 Gy five days per week. Generally, the radiation should be administered through parallel opposed lateral fields. The dose to the posterior cervical nodes is limited to 45 Gy (PTV1). Spinal cord blocks are used at a dose of 40 to 45 Gy. An electron beam is often used to treat any of the posterior neck which may require treatment to a higher dose (PTV4). A single anterior "supraclavicular" field is used to treat low neck nodes, supraclavicular, infraclavicular and upper mediastinal nodes when these nodes are at sufficient risk to harbor microscopic disease (PTV2). Treatment is usually prescribed to treat at a depth of 3 cm and the supraclavicular field is matched to the lateral fields on the skin. Care must be taken to include a block throughout treatment over the lower spinal cord in the lateral field since the supraclavicular field diverges into the lateral fields. Table 2 is a summary of the four PTVs used in treating the larynx, the doses delivered to them, and the typical treatment techniques used.

Treatment Verification

With the 2 mm cuts as identified in our CT scan localization/optimization plan, the DRRs should provide sufficient resolution to identify structures such as the cricoid cartilage, thyroid cartilage, mandible, aerodigestive tract and vertebral bodies.

CASE ILLUSTRATION

Presently CT simulation has limited value in the treatment of early larynx cancers since cancers can be relatively well localized on conventional simulation films. More advanced

Table 2. Summary of PTVs, Dose Delivered and Treatment Beam Arrangements Used to Treat the Four Volumes

PTV	Dose (Gy)	Treatment Technique
PTV1	40 to 45	Parallel opposed lateral fields
PTV2	45 to 50	En face photon field
PTV3	40-45 to 66-70	Parallel opposed lateral fields (exclude spinal cord)
PTV4	40-45 to 45-60	Lateral electron field

laryngeal cancer may benefit from the use of CT simulation for improved tumor localization, field design, and to develop novel treatment approaches. However, the quality of DRRs in this particular site are inferior to conventional radiographs. Therefore, conventional simulation films are often required following CT simulation and treatment planning for laryngeal cancer. A case is presented to illustrate the use of CT simulation in laryngeal cancer:

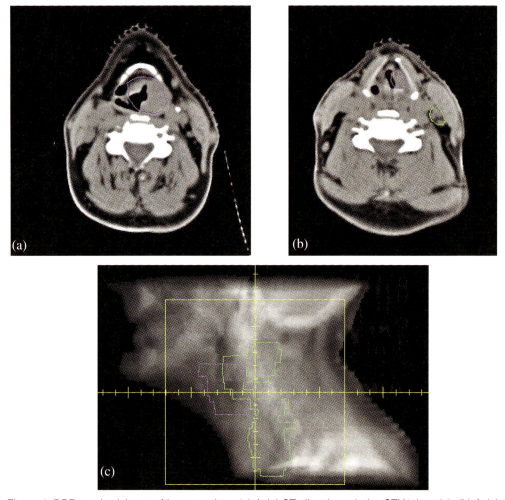

Figure 4. DRRs and axial cuts of larynx patient. (a) Axial CT slice through the CTV1 (purple). (b) Axial CT slice through CTV1 and involved node (green). (c) Left lateral DRR showing CTV1 and involved nodes.

The patient was a 65-year-old male who presented with dysphagia and a change in voice quality. Examination revealed a lesion of the left aryepiglottic fold that spilled into the medial wall of the pyriform sinus. The left true vocal cord had reduced mobility. The left neck had a palpable 4×3 cm jugulodigastric node and a 4×5 cm low jugular node. The right neck contained a 3×2 cm low jugular node. Biopsy revealed squamous cell carcinoma and he was staged a T3N2 squamous cell carcinoma of the aryepiglottic fold. He was referred for definitive treatment with radiation therapy alone. He received 44 Gy in 2 Gy fractions to the primary and upper necks via opposed lateral photon fields. After the administration of 44 Gy a spinal cord block was placed and the volume was boosted to 70 Gy. Following placement of the spinal cord block the bilateral posterior cervical nodes were boosted with 9 MeV electrons for an additional 6 Gy. The supraclavicular nodes received 50 Gy in 2 Gy fractions via an en face photon field. Figure 4(a) is an axial CT slice through the center of CTV1 (purple) while (b) shows another slice through the CTV1 and an involved lymph node (green). Figure 4(c) is a left lateral DRR showing CTV1 and involved nodes. The slice thickness and spacing were 5 mm and 3 mm, respectively.

PAROTID GLAND

INTRODUCTION

The parotid gland is the largest of the major salivary glands and is the one most frequently involved with benign and malignant tumors. Along with the paired submandibular and sublingual glands, the parotid gland comprises the major salivary glands.

ANATOMY

The superior border of the parotid gland is the zygomatic arch. The inferior border is the upper border of the posterior belly of the digastric muscle. The anterior border may extend to the orifice of the parotid duct opposite the second molar tooth. The parotid lies below and anterior to the external auditory canal and mastoid tip. Eighty percent of the gland lies on the outer surface of the masseter muscle, ascending ramus and angle of the mandible. The remainder of the gland is situated between the posterior edge of the mandibular ramus and the anterior borders of the sternocleidomastoid muscle as well as the posterior belly of the digastric muscle. This portion, which represents approximately 20% of the total mass of the gland, is called the tail of the parotid. The gland weighs between 14 and 28 grams. Contained within the gland are the facial nerve, part of the auriculotemporal nerve, part of the carotid artery as well as its terminal branches, the maxillary and superficial temporal arteries, the posterior facial vein and intraparotid nodes. It is a common misconception that the parotid gland is divided anatomically into superficial and deep lobes. Rather, the surgeon artificially divides the gland into superficial and deep lobes utilizing the facial nerve as the demarcation point. The main parotid duct (Stenson's duct) demarcates the anterior aspect of the parotid gland; it pierces the buccal mucosa and opens intraorally opposite the second upper molar. The sensory nerve is the greater auricular nerve. The superficial temporal and maxillary veins combine within the parotid gland to form the retromandibular vein. The arterial supply is provided by branches of the external carotid artery, the external facial, transverse facial, and occipital arteries.[1,2,7] Figure 5 is a schematic of the anatomy of the parotid gland.

STAGING

T1:   tumor 2 cm or less in greatest dimension.
T2:   tumor more than 2 cm but not more than 4 cm.
T3:   tumor more than 4 cm but not more than 6 cm.
T4:   tumor more than 6 cm.

All categories are subdivided into A) no local extension, and B) local extension. Local extension is defined as invasion of skin, soft tissues, bone, or nerve.

PATTERNS OF SPREAD

Parotid tumors are divided into benign and malignant tumors. In large part, this dictates their patterns of spread as well as the subsequent treatment. The benign tumors tend to grow slowly and rarely involve skin, bone, or nerve. On the other hand, the malignant neoplasms may deeply infiltrate the parotid gland and invade the facial nerve, skin, muscle, and bone. Lesions involved in the deep lobe may involve the base of the skull as well as the parapharyngeal space.

Lymphatic Spread

The incidence of lymph node metastasis is dependent upon the grade of the lesion, the AJCC stage, recurrent vs. primary lesion and the size of the lesion. A review of the Memorial Sloan-Kettering experience by Armstrong et al. revealed that 60 of 386 patients (16%) presenting with parotid tumors had clinically positive lymphadenopathy.[8] In addition, 30 of 326 patients (9%) had occult nodal metastasis. Multivariate analysis revealed only size and grade were significant risk factors for the occult nodal metastases when all major salivary gland tumors were included. Tumors 4 cm or greater in size had a 20% risk of occult metastasis compared with a 4% risk in smaller tumors. In addition, high grade tumors had a 49% risk of occult metastases compared with 7% in intermediate grade and low grade tumors.

Pathology

Parotid masses can be divided into benign and malignant lesions. Approximately 80% of primary lesions involving the parotid gland are benign. Benign lesions include: 1) benign mixed tumors which are the most common tumor of the salivary gland; 2) Warthin's tumor which primarily occurs in the tail of the parotid and is bilateral in 10% of cases; 3) Godwin's tumor which occurs more often in women and may be related to the HIV infection; 4) oncocytoma; and 5) basal cell adenoma.

Malignant tumors can be divided into low grade and high grade tumors. The low grade lesions include: 1) acinic cell carcinoma which is more common in women and notorious for recurring several decades after surgical resection; and 2) mucoepidermoid carcinoma, low grade.

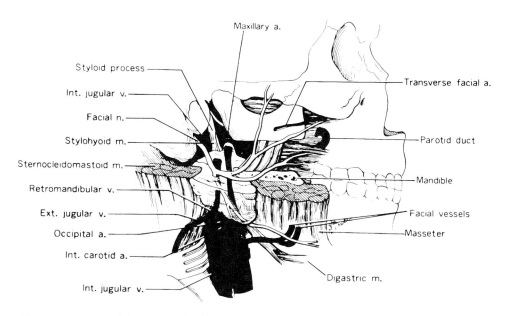

Figure 5. Anatomy of the parotid gland.[1]

The high-grade malignant lesions can be divided into: 1) mucoepidermoid carcinoma, high grade which is characterized by a high incidence of lymph node invasion as well as distant metastases; 2) adenocarcinoma, poorly differentiated carcinoma, anaplastic carcinoma and squamous cell carcinoma which are extremely aggressive tumors with a poor prognosis; 3) malignant mixed tumors; and 4) adenoid cystic carcinoma which is characterized by perineural invasion as well as recurrence many years following surgical resection.

Diagnostic Studies

CT and MRI are the examinations of choice for evaluating mass lesions within the parotid gland. Information provided by the CT or MRI should include: 1) the presence or absence of a mass; 2) whether it is marginated or infiltrating; 3) whether the mass is confined to the gland or has extended outside the capsule; 4) the presence of intraparotid or neck lymphadenopathy; 5) the presence of unilateral or bilateral disease; 6) the presence of multifocal disease within the parotid; 7) whether the mass is cystic, solid, or necrotic; and 8) the position of the mass relative to the facial nerve.[1,2]

IMMOBILIZATION

Patient Positioning

All patients should be simulated and treated in the supine position utilizing a headholder and immobilization device (i.e., either an Aquaplast mask or a bite block). It is important to ensure that either device is constructed from materials that produce little if any streaking artifact on the axial CT data. High density polystyrene is an excellent material for this purpose as it provides rigidity with little or no streaking artifact. Devices containing any form of metal should be avoided.

LOCALIZATION

Gross Tumor Volume/Clinical Tumor Volume

By convention the GTV is equated with findings on the CT or physical examination. Findings on clinical evaluation are correlated with information derived from axial CT or MRI data. As a result, at the Fox Chase Cancer Center the CTV is defined at the time of CT simulation on the axial CT data and the GTV is encompassed in this volume. There are a total of seven CTVs that are typically treated and are listed below.

CTV 1. Involved parotid bed
CTV 2. Cervical nodes
CTV 3. Subdigastric nodes
CTV 4. Posterior cervical nodes
CTV 5. Supraclavicular node
CTV 6. Upper mediastinal nodes
CTV 7. Infraclavicular nodes

Planning Target Volume

The PTV should include the CTV plus a margin for organ motion and set up uncertainty. Previous studies have reported set up uncertainties (one standard deviation) of 1.5 mm,[3] 2.1 mm,[4] and 4.0 mm[5] for head and neck treatments. For 95% coverage, a margin of two standard deviations around the CTV is used to define the PTV. Using these figures, margins can accordingly range from 3.0 mm to 8.0 mm. Two PTVs are used for treatment of the parotid gland and are summarized in Table 2 below. If the tumor is >4 cm, high grade, or if lymph nodes are involved, two PTVs are treated, otherwise a single PTV is treated for the entire course of radiation. PTV1 and PTV2 are defined below.

PTV1 = CTV1 + CTV2 + CTV3 + CTV4 + 3.0-8.0 mm margin
PTV2 = CTV5 + CTV6 + CTV7 + 3.0-8.0 mm margin

Definition of Clinical Structures

Various normal structures are typically included within the irradiated volume as a result of selected beam geometry and proximity to the tumor. Tolerances of normal tissues irradiated can be defined by their TD 5/5 which gives the maximal dose that can be delivered to the organ with minimal (5%) risk of complications. Table 3 lists the various normal tissues at risk in the treatment of parotid tumors and their respective TD 5/5 values.[6]

CT Scan Localization/Optimization & Virtual Simulation

Patient Position
The patient should be positioned with his/her head in a neutral position. Similar to larynx treatments, a neutral position ensures that organs are dissected perpendicularly to the CT slice. This reduces the possibility of oblique dissection of anatomical information and hence distortion.

Scanning Borders

Scanning borders are, superiorly, 2 cm superior to the zygomatic arch, and inferiorly, the cricoid cartilage, unless the supraclavicular fossa is irradiated in which case the inferior border is the top of the aortic arch.

Image Technique

CT scans are performed using predefined head and neck protocols stored on the CT scanner and bone windows.

Slice Protocol

Images should be obtained at 2 mm thicknesses with increments of 2 mm throughout the scan volume. This ensures optimal resolution for DRR images.

Contrast

100 to 150 cc of iodinated intravenous contrast should be administered by bolus and drip infusion immediately before and during the scan in order to help differentiate normal vessels from nodes, visualize the margins of the tumors with increased clarity, and determine the vascularity of the tumor.

Localization and Virtual Simulation

Treatment localization and CT simulation procedures are performed by the therapist using the guidelines outlined in Chapter 3. For the parotid, the treatment isocenter is defined as the center of CTV1 as outlined by the physician. Other structures such as CTV2- CTV7, and critical normal ones such as spinal cord are outlined by the physician and/or dosimetrist. Beam arrangement (virtual simulation) is then performed by the dosimetrist and physician. DRRs of the planned fields are regenerated and conformal blocks added.

Table 3. Normal Tissue Tolerances (Gy) and Resultant Complications When Treating Parotid Gland

Organ	TD 5/5 (Gy)	Complication
Parotid	32	Xerostomia
Soft tissue/bone	60	Bone exposure/ osteoradionecrosis
Temporal lobe	60	Necrosis (<1/3 volume)
	50	Necrosis (<2/3 volume)
	45	Necrosis (total volume)
Spinal cord	45	Myelitis
Skin	70	Fibrosis (<10 cm^2)
	60	Fibrosis (<30 cm^2)
	55	Fibrosis (<100 cm^2)
External auditory meatus	30	Acute serous otitis
	55	Chronic serous otitis

TREATMENT PLANNING

Radiation therapy to the parotid gland may be administered via an ipsilateral en face mixed beam approach (a combination of electrons and photons) or by a wedge pair technique. The direction of the wedge pair portals may be either superiorly and inferiorly directed or anteriorly and posteriorly directed as long as irradiation of the contralateral orbit is avoided. The primary advantage of the wedge pair technique is reduced dose to deep tissues, i.e., rapid dose fall-off. If the clinician carefully weights the electron and photon fields in the ipsilateral technique then the dose to the contralateral parotid should be <30 Gy, thus avoiding permanent xerostomia. Advantages of the en face mixed beam technique include the relative ease of matching a low neck field to the primary portal and the ability to shape the field with cerrobend blocks. Most parotid tumors are irradiated postoperatively utilizing fraction sizes of 1.8 to 2.0 Gy. The total dose to the CTV should be between 50 Gy and 60 Gy depending on the risk factors. Table 4 summarizes the two PTVs and doses used to treat the parotid gland.

TREATMENT VERIFICATION

With 2 mm cuts the DRRs will provide sufficient resolution to identify the vertebral bodies, orbits, temporal lobe, brainstem, mandible, thyroid cartilage, and aerodigestive tracts. In the treatment of parotid tumors the DRR achieves sufficient resolution so conventional simulation is not required.

CASE ILLUSTRATIONS

The first case involved a 46-year-old male who underwent a left superficial parotidectomy in 1968 for pleomorphic adenoma of the left parotid. In April 1993 he developed a 3 cm recurrence within the left parotid bed as well as a 7 mm left supraclavicular node. He underwent parotidectomy with preservation of the facial nerve in July 1993. Pathology revealed recurrent pleomorphic adenoma with a positive margin of resection. In October 1993 he developed another recurrence in the operative bed with pathology revealing pleomorphic adenoma. In February 1994 he again underwent excision of the recurrent disease with preservation of the facial nerve. Tumor was again left in situ and the patient was referred for postoperative irradiation. The patient underwent CT simulation and treatment planning. He received a total of 66 Gy to the parotid bed and ipsilateral neck using 16 MeV electrons and 6 MV photons in a 4:1 ratio. After the administration of 40 Gy a spinal cord block was used and an additional 26 Gy was given to the posterior cervical nodes using 9 Mev electrons. The left supraclavicular fossa was irradiated using 6 MV photons. The contralateral parotid gland received <4 Gy. Figure 6a shows an axial CT slice with the involved parotid bed (CTV1) outlined in purple while Figure 6b is a left lateral DRR showing extent of CTV1.

The second case involved a 50-year-old male who developed left parotid swelling in 1993 which was presumed to be a sebaceous cyst. He underwent local excision of the mass with pathology revealing carcinoma of parotid origin. Margins of resection were positive. He subsequently underwent a left superficial parotidectomy with pathology revealing adenocarcinoma with a close margin of resection. Lymph node sampling was negative. Due to the close margin a course of postoperative irradiation was recommended. He received 60 Gy to the left

Table 4. PTVs Used to Treat the Parotid Gland and Involved Lymphatic Nodes

PTV	Dose (Gy)	Treatment Criterion
PTV1	60-70	All patients

PTV2	45-50 to 60-70	If tumor:
		    >4cm
		    High grade
		    Lymph nodes involved

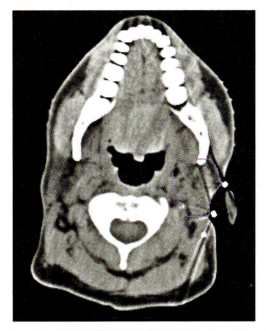

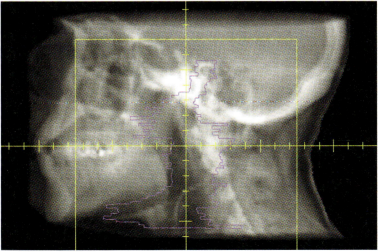

Figure 6. Axial CT slice and DRR for parotid patient #1. (a) Axial CT slice showing CTV1 (purple). (b) Lateral DRR showing extent of disease.

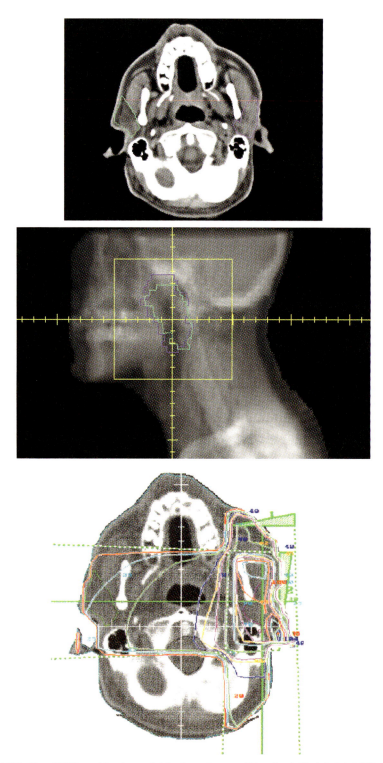

Figure 7. Axial CT slice, DRR, and isodose distributions for parotid patient #2. (a) Axial CT slice showing involved (purple) and normal (green) parotid. (b) Left lateral DRR showing involved and normal parotids. (c) Axial CT slice showing wedge pair arrangement used to treat parotid and isodose lines. Isodose lines have been normalized so that 100% represents 100% of the daily prescription.

parotid bed utilizing a wedge pair technique with 6 MeV electrons and 6 MV photons weighted 4:1. Figure 7a is an axial CT slice taken through the center of the CTV1 showing the involved (purple) and contralateral normal (green) parotid. Figure 7b is a lateral DRR showing the relationship between the normal and involved parotids while Figure 7c is an axial CT slice showing the wedge pair beam arrangement and isodose lines used to treat the patient. The display has been normalized so that the 100% isodose line (red) represents 100% of the daily prescription.

REFERENCES

1. Million RR, Cassisi NJ, Mancuso AA. Management of Head and Neck Cancer: A Multidisciplinary Approach. Philadelphia, Pennsylvania: J.B. Lippincott, 1994.

2. Som PM. Head and Neck Imaging. 2nd edition. St. Louis, Missouri: Mosby Year Book, 1991.

3. Hunt MA, Kutcher GJ, Burman C, Fass D, Harrison L, Leibel S, Fuks Z. The effect of setup uncertainties on the treatment of nasopharynx cancer. Int J Radiat Oncol Biol Phys 1993; 27: 437-447.

4. Dunscombe PB, Fox K, Loose S, Leszczynski K. The investigation and rectification of field placement errors in the delivery of complex head and neck fields. Int J Radiat Oncol Biol Phys 1993;26:155-161.

5. Rosenthal SA, Galvin JM, Goldwein JW, Smith AR, Blitzer PH. Improved methods for determination of variability in patient positioning for radiation therapy using simulation and serial portal film measurements. Int J Radiat Oncol Biol Phys 1992; 23:621-625.

6. Emami B, Lyman J, Brown A. Tolerance of normal tissue to therapeutic irradiation. Int J Radiat Oncol Biol Phys 1991;21:109-122.

7. Simpson JR. Principles and Practice of Radiation Oncology. 2nd edition. Philadelphia, Pennsylvania: J.B. Lippincott, 1992.

8. Armstrong JG, Harrison LB, Thaler HT, Friedlander-Klar H, Fass DE, Zelefsky MJ, Shah JP, Strong EW, Spiro RH. The indications for elective treatment of the neck in cancer of the major salivary glands. Cancer 1992; 69:615-619.

Chapter 9

PANCREATIC CANCER

Özer Algan, MD, John Hoffman, MD, and Andrew Shaer, MD

INTRODUCTION

Pancreatic cancer continues to remain an aggressive, debilitating, and fatal disease. Approximately 24,000 new cases of pancreatic cancer and approximately 27,000 deaths from pancreatic cancer are estimated for 1995, thus making pancreatic cancer the fifth most common cause of cancer related mortality.[1] At the Fox Chase Cancer Center, the majority of patients with pancreatic cancer have been treated with preoperative chemoradiation followed by surgical resection,[2] and this chapter will mainly focus on CT simulation for preoperative treatment. Although CT simulation and conformal treatment play an important role in the postoperative treatment of pancreatic cancer as well, its role in this setting is mainly limited to determining the location of normal anatomy, and to a degree, the gross tumor bed.

ANATOMY

The pancreas is a retroperitoneal organ that lies transversely in the posterior peritoneum of the upper abdomen. Grossly, it is a long, finely lobulated organ that extends from the right side of the L1-3 vertebral bodies across the midline towards the splenic hilum. Important relationships between the pancreas and its surrounding organs are shown in Figure 1. Going from right to left, the pancreas is anatomically divided into the head, neck, body, and tail. The head of the pancreas sits in the curve of the duodenum forming the "C" loop. The first part of the duodenum can be superior or anterior to the head of the pancreas. The posterior surface of the head lies on the right renal hilum and its associated vasculature. The common bile duct passes through the upper part of the pancreatic head as it extends toward the second part of the duodenum. The uncinate process of the pancreas lies inferior to the neck and is anterior to the aorta. The superior mesenteric vessels pass posterior to the neck but anterior to the uncinate process (Figure 2). The neck lies anterior to the aorta and the superior mesenteric vessels. The body lies superior to the duodenojejunal flexure and posterior to the lesser sac. It crosses the left kidney just above the hilum and tapers into the tail. The tail follows the splenic vasculature all the way to the splenic hilum.[3] For staging purposes, the pancreas is divided into the head, body, and tail. Pancreatic head tumors are those arising to the right of the left border of the superior mesenteric vein. The uncinate process is considered a part of the pancreatic head. Tumors of the pancreatic body are those arising between the left border of the superior mesen-

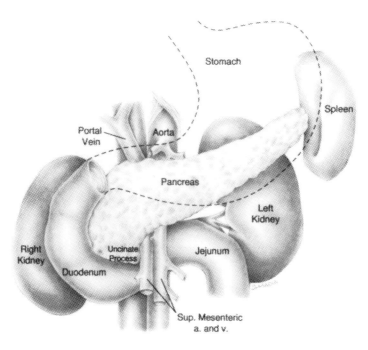

Figure 1. Normal anatomy of the pancreas and surrounding tissues.

teric vein and the left border of the aorta. Tumors of the pancreatic tail are those arising to the left of the left border of the aorta.[4]

The main pancreatic duct extends from the tail, through the body to the posterior inferior portion of the neck and head, including the uncinate process. In most cases it joins the common bile duct as it drains into the second part of the duodenum. The accessory pancreatic duct drains the superior part of the pancreatic head and empties into the second part of the duodenum. The celiac trunk originates at the upper margin of the pancreas and gives off the common hepatic artery, the splenic artery, and the left gastric artery. The portal vein, which is formed by the confluence of the splenic vein and the superior mesenteric vein, lies posterior to the pancreatic head.[3]

The vasculature and the lymphatic drainage of the pancreas are shown in Figure 2. The pancreas is supplied by arterial branches of the celiac trunk and the superior mesenteric artery. The venous drainage generally follows the arterial supply, and eventually empties into the portal vein and the superior mesenteric vein. The peripancreatic lymphatics (including the superior, inferior, anterior, and posterior peripancreatic lymph nodes) form a rich interanastomosing network as they run along the surface of the pancreas. Lymphatic drainage from the head of the pancreas predominantly goes to the celiac or the superior mesenteric lymph node chains. However, the peripancreatic (including the superior, inferior, anterior, and posterior peripancreatic), hepatic, infrapyloric, subpyloric, celiac, superior mesenteric, retroperitoneal, and lateral aortic lymph node chains are all considered regional lymph nodes.[4] Rarely, drainage into the lumbar lymph nodes can also occur. The body and the tail of the pancreas drain into the pancreaticolienal lymph nodes along the splenic vessels into the splenic lymph node chain as well as the inferior and superior peripancreatic lymph nodes of the pancreatic body and tail.[5] For pancreatic tail and body lesions, the peripancreatic, hepatic, superior mesenteric, pancreaticolienal, splenic, retroperitoneal, and lateral aortic lymph node chains are all considered regional nodes.[4] Posterior lesions also have a propensity to drain into the paraaortic lymph nodes.

PATTERNS OF SPREAD

The majority of pancreatic tumors are adenocarcinomas, with approximately two thirds of them originating in the head of the pancreas. The majority of patients present with advanced disease where local-regional spread is a common finding. For lesions involving the pancreat-

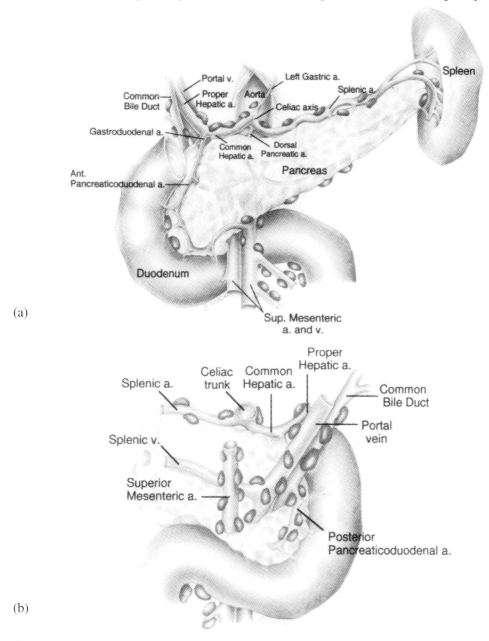

Figure 2. Vascular and lymphatic drainage of the pancreas (a) from an anterior view, and (b) from a posterior view.

ic head, involvement of adjacent organs occurs in 25% to 50% of all patients.[6,7] Obstructive jaundice is a common finding. Lesions in the body and tail of the pancreas are usually larger at presentation and tend to involve the splenic vasculature. Involvement of other vasculature such as the portal venous system, superior mesenteric artery and vein or peritoneal seeding is not unusual. Lymph nodes that are at highest risk for spread of disease from pancreatic head cancers include the posterior pancreaticoduodenal lymph nodes in 47% to 56% of patients, lymph nodes superior to the pancreatic head in 22% to 48%, lymph nodes superior to the pancreatic body in 28%, lymph nodes inferior to the pancreatic head in 24%, anterior pancreaticoduodenal lymph nodes in 13% to 47%, and lymph nodes inferior to the body in approximately 1% of patients.[5,8] Because of the lower incidence of cancer in the pancreatic body and tail, less information is known about these lesions. However, a retrospective study with limited number of patients done by Nagai et al.[9] suggests involvement of the lymph nodes superior to the pancreatic body in 66% (2 of 3 patients), and the paraaortic lymph nodes in 33% (1 of 3 patients) of patients with cancers in the pancreatic body or tail. Distant metastasis to the liver can occur early and is found in 15% to 25% of patients at presentation.[6] Other sites of distant metastases include the peritoneal cavity, lung, pleura, and rarely the brain. Overall, less than 10% to 20% of patients at diagnosis will have microscopic disease limited to the pancreas.[7,10] Approximately 40% of patients present with locally advanced disease with involvement of the regional lymph nodes or adjacent pancreatic structures, and 40% will have identifiable visceral metastases.[10] Peritoneal seeding occurs in approximately 35% of patients. The pattern of failure reflects the locally advanced nature of the disease with approximately 45% to 53% having local-regional failure, 30% to 44% liver metastases, 31% to 40% peritoneal seeding, and 20% distant metastases.[7,11,12] Hepatic failure either from obstructive jaundice or hepatic metastases remains the major cause of death.

DIAGNOSTIC STUDIES

Computed tomography scan using venous portography with late portal views remains the best study for evaluating the extent of disease. It allows for the visualization of the pancreatic mass, any grossly enlarged peripancreatic lymph nodes, the surrounding vasculature, the liver, and the peritoneal cavity for ascites. A helical or spiral CT scanner is better at visualizing the peripancreatic lymph nodes, and provides an excellent new means for evaluating vascular involvement. Other studies should include transhepatic cholangiography or endoscopic retrograde cholangiopancreatography to evaluate the location and extent of biliary tree involvement as well as to palliate the obstructive jaundice, and a chest x-ray to rule out pulmonary metastasis. In the absence of spiral CT, if there is suggestion of vascular invasion or question about normal anatomy, angiography should be performed to further evaluate the superior mesenteric artery, the vessels of the celiac axis, and the portal system. Endoscopic ultrasound may also help determine portal vein involvement in questionable cases. Although preliminary retrospective studies have suggested improved preoperative staging with laparoscopy,[13] the role of laparoscopy for staging remains unclear and is currently being evaluated in prospective trials. The ability of MRI to evaluate primary pancreatic carcinoma remains to be seen.

IMMOBILIZATION

Patients are CT simulated and treated in the supine position with their arms overhead. For immobilization, a custom Alpha Cradle cast extending from the mid-thoracic spine to the mid-pelvis is used.

LOCALIZATION

Definition of Gross Tumor Volume

The gross tumor volume (GTV) represents tumor in the pancreas or the surrounding tissue that is detected by physical exam or more commonly by imaging studies. The clinical target volume (CTV) includes the GTV plus any microscopic extension of disease that is suspected around the GTV. For lesions involving the pancreatic head, the lymph nodes that are at highest risk for microscopic spread include the pancreaticoduodenal, the celiac, the porta hepatic, and the suprapancreatic and infrapancreatic lymph node chains. The duodenal loop as well as the peripancreatic fatty tissue are also at high risk for microscopic involvement. For lesions involving the body and the tail of the pancreas, the lateral suprapancreatic and the splenic hilar lymph nodes are at highest risk for microscopic spread. The planning target volume (PTV) represents the clinical target volume plus any additional margins that are required to compensate for daily set up variations or internal organ motion. At the Fox Chase Cancer Center, a 1 cm margin is used around the CTV to determine the PTV. However, it should be noted that the inclusion of all of the above mentioned tissue can result in fairly large fields, and it may not be possible to achieve these margins in certain areas, particularly in the region of the kidneys and the spinal cord. Therefore, treatment planning needs to be individualized to each patient based on the location and size of the tumor as well as the location of the surrounding critical structures.

Definition of Critical Structures

The critical structures of concern during the treatment of pancreatic cancer include the stomach with a maximum tolerable dose of 60 Gy, the small intestine with a maximum tolerable dose of 50 Gy, the liver with a maximum tolerable dose of 50 Gy, the kidneys with a maximum tolerable dose of 50 Gy, and the spinal cord with a maximum tolerable dose of 50 Gy (Table 1). All of these figures represent the TD 5/5 to one-third of the target volume as described by Emami et al.[14] Concurrent chemotherapy or previous abdominal surgery may further decrease the tolerance of these tissues. The liver, kidneys, and the spinal cord are the usual dose limiting structures in the treatment of pancreatic cancer and special attention must be paid to the amount of radiation that is received by these organs during treatment planning.

CT SCAN LOCALIZATION AND OPTIMIZATION

Patients are given 10 ounces of Readi-CAT[1] oral CT contrast one hour prior to starting CT simulation to opacify the gastrointestinal tract, and 5 ounces of Readi-CAT[1] oral contrast

Table 1. Tissue Dose Tolerance (TD 5/5) for Irradiation of One-Third and Two-Thirds of Target Volume*

Target Organ	TD 5/5 to 1/3 of the Target Volume	TD 5/5 to 2/3 of the Target Volume
Stomach	60 Gy	55 Gy
Small intestine	50 Gy	NS†
Liver	50 Gy	35 Gy
Kidney	50 Gy	30 Gy
Spinal cord	50 Gy	50 Gy

*Modified from Emami et al., Int J Radiat Oncol Biol Phys, 21(1):109-121, 1991.
†NS: Not specified.

immediately prior to CT simulation to help distinguish the pancreas from the stomach and duodenum. Generally, intravenous contrast is not used unless there are specific concerns about the vascular structures. Each patient is scanned from the upper dome of the right diaphragm (approximately at the level of T9-T10) to the bottom of the L4 vertebral body. The patient is imaged with overlapping CT slices that are 5 mm thick at 3 mm intervals. This allows for good tumor and critical tissue visualization while minimizing the extent of volume averaging.

TREATMENT PLANNING

Dose Prescription for Clinical Target Volume and Critical Structures

Radiation therapy is usually given with concurrent chemotherapy in the treatment of pancreatic cancer. Treatment planning for radiation therapy must be individualized for each patient based on the size and location of the lesion, as well as the normal anatomy surrounding the lesion. As a general guide regarding the field borders of the initial field from conventional simulation principles, the superior margin should be at the superior border of T11 vertebral body to insure coverage of the celiac axis, the anterior margin is usually 1.5 to 2.0 cm anterior to the lesion, and the posterior margin is 1.5 cm behind the anterior portion of the vertebral body. In patients that are deemed unresectable, a large field is treated initially (PTV1) followed by a cone-down to the gross tumor (PTV2). The total tumor dose is 60 Gy to the gross disease (PTV2 = GTV + 1 cm), and 50 Gy to the microscopic disease (PTV1 = CTV + 1 cm). In patients that are treated preoperatively, the treatment dose is 50 Gy to PTV1 using 1.8 to 2.0 Gy fractions. Other goals of treatment planning include keeping the dose to one-half of one functioning kidney under 18 Gy, keeping the dose to one-half of the liver under 20 Gy, and keeping the dose to the spinal cord under 45 Gy. An effort must also be made to minimize the amount of bowel in the treatment field, as well as keeping the dose to segments of bowel greater than 5 cm under 50 Gy. Typically, multifield beam arrangements are necessary to meet the normal tissue constraints.

TREATMENT VERIFICATION

Digitally reconstructed radiographs (DRR) alone in general do not provide enough information regarding soft tissue anatomy in the abdomen. Therefore, at present all patients treated for pancreatic cancer undergo conventional simulation in addition to CT simulation at the Fox Chase Cancer Center. Subsequently, the target volume as well as other critical structures such as the spinal cord, kidneys, liver, stomach, and duodenum can be transferred from the CT treatment plan to the conventional simulation field by overlaying the two films. However, advances in image manipulation resulting in improved DRR images should eliminate the need for this step in the near future.

CASE STUDIES

Patient #1

The first patient is a 72-year-old female with severe scoliosis who presented to her family physician with several months history of dark urine, and several days history of loose, clay-colored stools and severe abdominal discomfort. CT scan of the abdomen revealed extrinsic compression of the common bile duct. ERCP revealed abnormal narrowing of the distal common bile duct. Transhepatic cholangiography with catheter placement revealed a distended gallbladder and a markedly dilated biliary tree. Pathology from the common bile duct was positive for an adenocarcinoma most consistent with a pancreatic primary. Ultrasound of the abdomen and liver revealed intrahepatic ductal dilation, cholelithiasis, and dilated proximal common bile duct. Arteriogram demonstrated external compression of the portal vein and the superior mesenteric vein. Chest x-ray was negative.

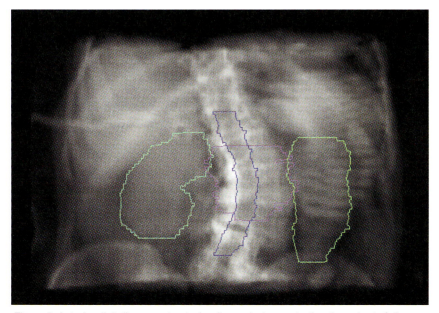

Figure 3. Anterior digitally reconstructed radiograph demonstrating the extent of disease as well as the normal anatomic structures. Green represents the kidneys, blue represents the spinal cord, and magenta represents clinical target volume. The stomach and the liver can also be seen. Also note the severe scoliosis in this patient.

Because of the vascular involvement, the patient was considered to be unresectable initially and was treated with preoperative radiation therapy with concurrent chemotherapy. Figure 3 is a DRR demonstrating the extent of disease as well as the normal anatomic structures. In this figure, the green wire frames represent the right kidney, the blue wire frame represents the spinal cord, and the magenta wire frame represents the CTV. The PTV was established by placing 1 cm margins around the CTV. The liver and the stomach can also be seen on the DRR. Because the mass was fairly midline, a plan was generated using initial anterioposterior fields to a dose of 41.4 Gy to the 95% isodose line. With this field arrangement, essentially all of the right kidney and greater than one-half of the left kidney were outside of the high dose area. Subsequently, the treatment portals were changed to exclude the spinal cord from the high dose area, and the patient received an additional 9 Gy to the 95% isodose line using a right anterior oblique, and a left lateral treatment portal to the same PTV. This allowed sparing of the spinal cord and the right kidney at the cost of increasing the dose to the left kidney. Figure 4 demonstrates the isodose distribution for the entire treatment plan on a transverse cut through the central axis of the treatment field. Overall, the spinal cord received less than 45 Gy, the right kidney received less than 10 Gy, and the left kidney received less than 50 Gy. The majority of the liver was outside of the treatment field or was outside of the high dose area. Also note the significant change in the patient's normal anatomy secondary to severe scoliosis. CT simulation is especially useful in this setting by allowing the exact localization of the tumor volume and the critical normal structures with in the abdomen.

Patient #2

The second patient is a 63-year-old female who presented with a 20-pound weight loss over three months, and a one month history of abdominal discomfort. CT scan of the abdomen revealed a large mass at the head of the pancreas with no evidence of metastasis. CT-guided

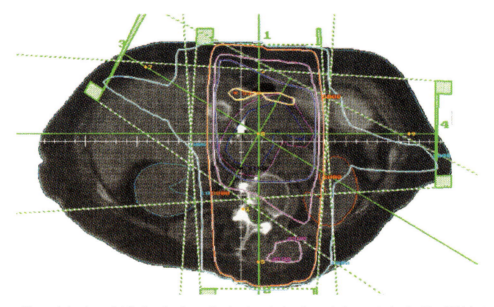

Figure 4. Isodose distribution for the entire treatment plan through the central axis. The PTV (outlined in magenta) is encompassed within the 95% isodose line and was treated to a total dose of 50.4 Gy. Overall, the spinal cord received less than 45 Gy, the right kidney received less then 10 Gy, and the left kidney received less than 50 Gy. Dark blue represents the CTV. The PTV is determined by adding a 1 cm margin around the CTV.

fine needle biopsy was positive for adenocarcinoma of the pancreas. Angiogram and CT portogram revealed vascular involvement. Laporascopic examination revealed liver metastases. The patient was referred to our department for consideration of palliative concurrent chemoradiation.

Figure 5 shows an anterior DRR with wire frames around the left and right kidney, the spinal cord, and the CTV. Figure 6 demonstrates the composite isodose distribution for the entire treatment plan on a transverse cut through the central axis. Because of the patient's right sided lesion, the right kidney was at risk for receiving significant radiation dose, and an effort was made to spare as much of the left kidney as possible. A plan was generated using a four-field arrangement with an anterior field, two right posterior oblique fields, and a left anterior oblique field. This allowed the PTV to be encompassed within the 95% isodose line, while allowing the spinal cord and especially the left kidney to receive significantly lower doses. The patient was treated to a total dose of 50 Gy to the 95% isodose line using 2 Gy fractions given daily. This resulted in the right kidney receiving approximately 50 Gy, the left kidney receiving less than 10 Gy, and the spinal cord receiving less than 45 Gy. This case demonstrates the importance of CT simulation in the development of complex treatment plans where knowing the exact location of the critical structures as well as the dose received by them is crucial. It would essentially be impossible to come up with a similar treatment plan using conventional simulation techniques secondary to the uncertainties in determining the exact location of the tumor volume as well as the critical structures.

CONCLUSION

CT simulation allows better localization of the target volume as well as the surrounding normal tissue than conventional simulation. This is especially true in the abdomen where it is

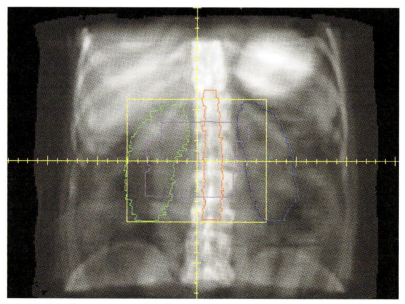

Figure 5. Anterior digitally reconstructed radiograph demonstrating the extent of disease as well as the normal anatomic structures. Green represents the right kidney, blue represents the left kidney, red represents the spinal cord, and magenta represents the clinical target volume.

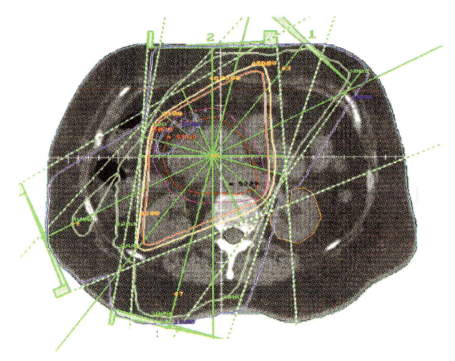

Figure 6. Composite isodose distribution for the entire treatment along the central axis of the clinical target volume. Using the above four-field arrangement, a total of 50 Gy was delivered to the 95% isodose line. This allowed significant sparing of dose to the spinal cord and especially to the left kidney. The CTV is shown in dark red. The PTV was determined by adding 1 cm around the CTV.

difficult to identify structures such as the liver, pancreas, and blood and lymphatic vessels on standard simulation films. The improved localization of important structures within the abdomen allows for field arrangements and treatment planning that delivers the maximum radiation dose to the target volume while minimizing the dose to the surrounding tissue. This allows the radiation oncologist to confidently treat the pancreatic carcinoma with field arrangements which are otherwise impossible with conventional simulation only. This would, in turn, be expected to decrease the treatment-related complications, while insuring the target volume receives the prescribed dose. The use of spiral CT simulators coupled with image-enhanced DRRs may obviate the need to do conventional simulation in addition to CT simulation in patients with pancreatic carcinoma.

REFERENCES

1. NCI-SEER Program. Cancer statistics 1994. CA-A Cancer Journal for Clinicians 44(1):7-26;1995.

2. Coia LR, Hoffman H, Scher R, Weese J, Solin L, Weiner L, Eisenberg B, Paul P, Hanks GE. Preoperative chemoradiation for adenocarcinoma of the pancreas and duodenum. Int J Radiat Oncol Biol Phys 30(1):161-167;1994.

3. The gut and its derivatives. In: Hollinshead WH, Cornelius R, eds. Textbook of Anatomy. Philadelphia, Pennsylvania: Harper & Row, 1985, pp 634-644.

4. American Joint Committee on Cancer Staging. Fourth edition. Philadelphia, Pennsylvania: J.B. Lippincott, 1994, pp 109-111.

5. Cubilla AL, Fortner J, Fitzgerald PJ. Lymph node involvement in carcinoma of the head of the pancreas area. Cancer 41:880-887;1978.

6. Howard JM, Jordan GL. Cancer of the pancreas. Curr Probl Cancer 2:1;1977.

7. Foo ML, Gunderson LL, Nagorney DM, McLlrath DC, VanHeerden JA, Robinow JS, Kvols LK, Garton GR, Martenson JA, Cha SS. Patterns of failure in grossly resected pancreatic ductal adenocarcinoma treated with adjuvant irradiation ± 5 fluorouracil. Int J Radiat Oncol Biol Phys 26:483-489;1993.

8. Ozaki H, Kishi K. Lymph node dissection in radical resection for carcinoma of the head of the pancreas and periampullary region. Jpn J Clin Oncol 13(2):371-378;1983.

9. Nagai H, Kuroda A, Morioka Y. Lymphatic and local spread of T1 and T2 pancreatic cancer. Ann Surg 204(1):65-71;1986.

10. Brooks J. Cancer of the pancreas. In: Brooks JR, ed. Surgery of the Pancreas. Philadelphia, Pennsylvania: WB Saunders, 1983, p 263.

11. Griffin JF, Smalley SR, Jewell W, Paradelo JC, Reymond RD, Hassanein RES, Evans RG. Patterns of failure following curative resection of pancreatic carcinoma. Cancer 66:56;1990.

12. Tepper JE, Nardi GL, Suit HD. Carcinoma of the pancreas: review of MGH experience from 1963 to 1973: analysis of surgical failure and implications for radiation therapy. Cancer 37:1519;1976.

13. Warshaw AL, Gu Z, Wittenberg J, Waltman AC. Preoperative staging and assessment of resectability of pancreatic cancer. Arch Surg 125:230-233;1990.

14. Emami B, Lyman J, Brown A, Coia LR, Goitein M, Munzenrider JE, Shank B, Solin LJ, Wesson M. Tolerance of normal tissue to therapeutic irradiation. Int J Radiol Oncol Biol Phys, 21(1):109-121;1991.

Chapter 10

LUNG CANCER

Nicos Nicolaou, MBChB, FRCPC

INTRODUCTION

Lung cancer is the most common malignancy in the United States with 157,000 new cases diagnosed annually. It accounts for one of every six new cancer cases and is responsible for one of every four deaths due to cancer. It is now the most common cause of cancer death in men and women causing 142,000 deaths annually. Long term survival has increased slightly with the overall five-year survival rate for all patients presently at 13% compared to 7% in 1963. Only 21% of patients have localized disease at diagnosis. Current diagnostic studies define approximately 20 to 25% of patients to be surgical candidates. Twenty percent of lung cancer patients will have small cell (SCLC), and the rest non-small cell (NSCLC) histology. The individual biology and the management of these two groups are quite different.[1,2]

Clinical stage (CS) I NSCLC is treated by complete surgical resection. Five year disease-free survival for resected pathologic stage (PS) I NSCLC is 50% to 80%. Definitive irradiation (RT) can be used for CS I NSCLC if surgery is not possible.[1] Optimum management of stage II cases begins with surgical staging and resection. Standard postoperative adjuvant therapy of PS II NSCLC consists of mediastinal and hilar RT. A lung cancer study group (LCSG) phase III trial compared postoperative adjuvant RT to observation in PS II and III squamous cell carcinoma patients and showed a statistically significant lower intrathoracic failure rate among the irradiated patients (3% vs. 19%, $P = <.001$).[3] Results of CS II NSCLC treated with RT are limited. Management of stage III NSCLC patients is under active reevaluation due to the disappointing locoregional control rates and poor long term survival. Clinical stage III NSCLC patients have a median survival of 9 to 13 months and a 10% to 20% two-year survival. Patients with no adverse prognostic features have a two-year survival rate of 20% to 30%.[2]

Phase III trials have compared irradiation with pre- and postchemotherapy vs. irradiation alone for unresected stage III NSCLC. The addition of Cisplatinum-containing chemotherapy to thoracic RT is supported by a survival advantage observed in three randomized trials. Induction chemotherapy (Vinblastine/Cisplatin) preceded thoracic RT in CALGB trial 84-33.[4] EORTC trial 08844[5] delivered weekly or daily Cisplatin concurrently with thoracic RT. The French multicenter trial CEBI 138[6] delivered induction and post RT chemotherapy (Vindesine/Cisplatin/Lomustine/Cyclophosphamide). The improvement in survival rate over RT alone in the trials using induction chemotherapy appears to be related to a decrease in detectable distant metastases, while the observed survival advantage by EORTC with low dose daily Cisplatin during thoracic RT was associated with improved control of locoregional disease. The simultaneous delivery of Cisplatin and RT presumably enhanced the local disease response and the use of higher drug doses in induction regimens deterred progression of micrometastatic disease.

The resection rate of selected patients with visible mediastinal adenopathy on chest x-ray is only 10% to 15%. Efforts to improve operability have included pre-operative chemotherapy ± RT. Some institutions report a resection rate of 50% to 70% with encouraging long term survival rates. Stage III completely resected PS IIIA patients are treated with postoperative RT. Local control and relapse-free survival advantages were observed among 33% of patients with stage IIIA disease in favor of RT in the randomized LCSG trial for resected stages II and IIIA squamous cell carcinoma.[4] There was no overall survival difference seen in another LCSG trial which randomized patients with incompletely resected stage III lesions to RT ± combined chemotherapy. The recurrence rate, however, was lower with combined therapy.

SCLC spreads systemically very early. Standard therapy for limited stage SCLC is concurrent multiagent Cisplatinum-containing chemotherapy and thoracic irradiation. Trials have demonstrated a survival advantage for using thoracic RT with chemotherapy vs. chemotherapy alone, as well as for early delivery of concurrent RT over delayed RT.[7,8] Pilot data using accelerated fractionated RT have been encouraging. Extensive stage SCLC is treated with chemotherapy and extent of RT utilization depends on initial response. Disagreement exists over prophylactic cranial irradiation (PCI) for limited stage SCLC. Opponents cite the associated late neurotoxicity with an overall crude incidence of approximately 19% and the absence of a demonstrated statistically significant survival benefit.[9] Proponents feel the toxicity to be related to large RT doses per fraction and the PCI given concurrently with the chemotherapy. It is also felt that the morbidity of brain failure warrants reduction even in the absence of improved survival.[10] Recent European studies of PCI have shown a two- to three-fold reduction in brain relapse rates, no increased toxicity and the prospect of a 7% improvement in survival.[11] PCI will continue to be controversial until well conceived and conducted trials resolve the above issues.

RT is the modality mainly utilized for the palliation of lung cancer patients and for treating the multitude of complications that can arise from locoregional and systemic disease extent. The ability of RT to improve the quality of life of these patients is well documented.

Local control of intrathoracic NSCLC with definitive RT using standard fractionation and total doses of 60 to 65 Gy is unacceptably low. An RTOG study showed that higher doses of irradiation yield a better complete response rate locally and a lower intrathoracic failure rate, as well as improved three year survival in unresectable and medically inoperable NSCLC patients.[12] Local issues, previously considered secondary to the problem of distant metastases in SCLC, are now becoming more dominant in patterns of failure for limited stage SCLC, since survival rates are presently approaching 50% and 30%, at 2 and 5 years respectively.

Most local failures in modern RT series appear to be infield for both NSCLC and SCLC which may indicate inadequate dose delivery.[13] The need to deliver higher doses of RT, however, has to be balanced against the concomitant increase in normal tissue toxicity. Smaller treatment volumes and higher targeting accuracy are therefore required for safe escalation of RT doses. Three dimensional conformal radiation therapy (3D CRT) is a mode of high precision radiotherapy that plans and delivers radiation treatments that shape the prescription dose distribution to conform to the target in its entire 3D configuration. Multiple static coplanar and non-coplanar fields with conformal customized blocks are used for improved accuracy of target coverage and a decrease in the volume of the normal tissues treated. This results in less RT-related complications, and allows increased dose delivery to the target beyond levels that were before possible with 2D RT.[14]

Utilization of 3D CRT in lung carcinomas should therefore enhance local control of primary malignancy, by allowing higher doses of RT to be delivered, while at the same time decreasing RT related morbidity through improved targeting.[15] The lower incidence of local failure and the expected decrease in the subsequent possible development of distant metastasis, as well as the lower associated morbidity should result in an improvement in therapeutic ratios and ultimately survival rates.

LUNG ANATOMY[6]

The lungs lie in close proximity to various dose limiting structures including the heart and spinal cord. The intrathoracic organs experience continuous movement due to cardiac and respiratory motion. Consideration of these factors when planning and delivering RT is essential for an acceptable therapeutic ratio.

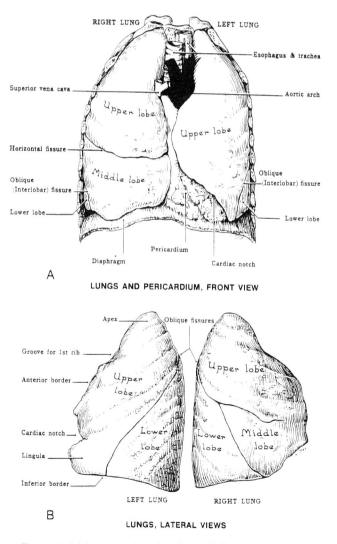

Figure 1. (a) Lungs and pericardium. (b) Lungs: posterolateral views.[6]

Gross Features

Each lung has a base, an apex, costal/medial surfaces, anterior/inferior borders and a hilus which is the entry point for the bronchial and neurovascular lung components. The lung base conforms to the diaphragmatic dome which ascends to mid-T8 on the right and T8/9 interspace on the left. The apex forms the cupola which rises above the first rib and clavicle under the fascia of the muscles of the neck. The apex is close to the brachial plexus and sympathetic stellate ganglion. The vertebral column and the mediastinum are between the medial surfaces of the right and left lungs. The vertebral aspect of the medial lung surface occupies the thoracic gutters on either side of the vertebrae. The mediastinal part contains the hilus or root of the lung which is bridged on the right by the azygous vein and on the left by the aortic arch. The right lung is separated into superior, middle, and inferior lobes by the horizontal and oblique fissures respectively. The left lung is separated into superior and inferior lobes by an oblique fissure. The lingula, a projection of the superior lobe, is analogous to the middle lobe of the right lung.

The mediastinum contains all the thoracic viscera except the lungs. Contents of the superior mediastinum include retrosternal structures (thymus, great veins), prevertebral structures (trachea, esophagus, left recurrent laryngeal nerve, thoracic duct) and intermediate structures (aortic arch, vagus, and phrenic nerves). The inferior mediastinum has three divisions each containing different structures, i.e., anterior (fat and thymic tissue), middle (pericardium, heart, great vessels) and posterior (descending aorta, thoracic duct, azygous/hemiazygous venous systems, esophagus).

Bronchial Tree

The midline trachea divides at the T4/5 interspace into right and left main (primary) bronchi which are contained in the mediastinum. They enter the hilum of each lung and divide into lobar (secondary) bronchi which extend into the lung tissue of each lobe. The right lung has three (upper, middle, lower) and the left lung two (upper, lower) lobar bronchi, which divide into segmental (tertiary) bronchi. The portion of the lung aerated by the subsequent division of the segmental bronchus is the bronchopulmonary segment (see Figure 2).

Vascular

The pulmonary arteries lie anterior to the main bronchi as they enter their respective lung hili. The azygous vein crosses the right pulmonary artery superiorly while the left pulmonary artery is situated inferiorly to the aortic arch at the level of T5. The pulmonary veins lie anterior to the pulmonary arteries and the bronchi, in the hilus of each lung.

Lymphatics

The regional lymph nodes are intrathoracic and extrathoracic.[17,18]
(a) Extrathoracic nodes: supraclavicular and scalene
(b) Intrathoracic nodes: mediastinal, peribronchial and hilar (see Figure 3).
(i) Mediastinal Subgroups
1   Superior mediastinal
2R   Right upper paratracheal   Between intersection of caudal margin of inominate artery

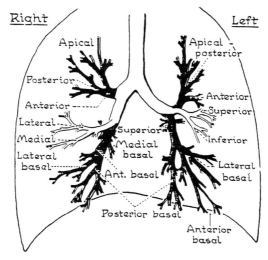

Figure 2. The distribution of the bronchi.[6]

		with trachea and the apex of the lung
2L	Left upper paratracheal	Between top of aortic arch and apex of the lung
3	Pretracheal	
3a	Anterior mediastinal	
3p	Posterior mediastinal	Retrotracheal
4R	Right lower paratracheal (Tracheobronchial)	Between intersection of caudal margin of inominate artery with trachea and cephalic border of azygous vein
4L	Left lower paratracheal (Tracheobronchial)	Between top of aortic arch and carina medial to ligamentum arteriosum
5	Aortopulmonary	Subaortic (aortic window) lateral to ligamentum arteriosum
6	Paraaortic	
7	Subcarinal	Caudal to the carina
8	Paraesophageal	Below carina to right or left of esophageal midline
9	Pulmonary ligament	Within right or left pulmonary ligament

(ii) Peribronchial Intrapulmonic and Hilar

10R	Hilar/main bronchus (Right peribronchial)	From level of cephalic border of azygous vein to origin of right upper lobe bronchus
10L	Hilar/main bronchus (Left peribronchial)	Between carina and left upper lobe bronchus medial to ligamentum arteriosum
11	Interlobar	
12	Lobar	Intrapulmonic
13	Segmental	

PATTERNS OF SPREAD

The spread of lung cancer can occur by local (intrathoracic), regional (lymphatic) and systemic (hematogenous) pathways. Primary malignancies metastasize to regional draining

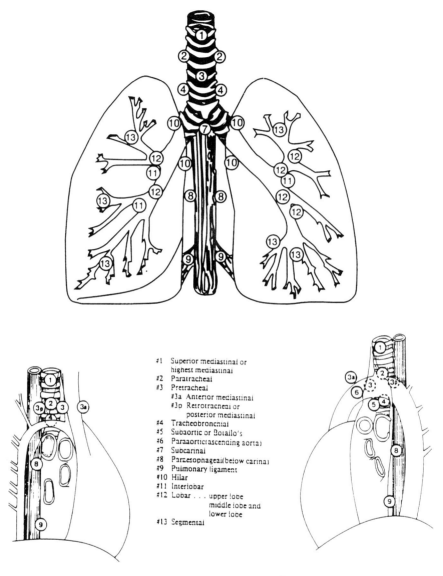

Figure 3. Intrathoracic regional lymph nodes of the lungs.[1]

lymph nodes and to distant sites after invading the local vascular and lymphatic channels. Lymphatic drainage in the lung follows the bronchoarterial branching pattern with lymph nodes situated at origin of these branchings. Lymphatic channels coalesce and drain into lymph nodes around segmental and lobar bronchi.[19] Rouviere (1932), Nohl-Osser (1972) and Spencer (1985) defined lymphatic pathways of malignant spread to mediastinum[20] (see Figure 4 and Table 1):

(a) Right Lung

Right upper lung lobe spreads to 10R, 4R, 2R (Figure 4a). Middle lobe to 10R, 4R, 2R, and 7 (Figure 4b). Right lower lobe to 10R, 4R, 2R, 7 (Figure 4c) and not infrequently to 8 and 9. In right lung malignancies contralateral involvement was found in only 4.1%.

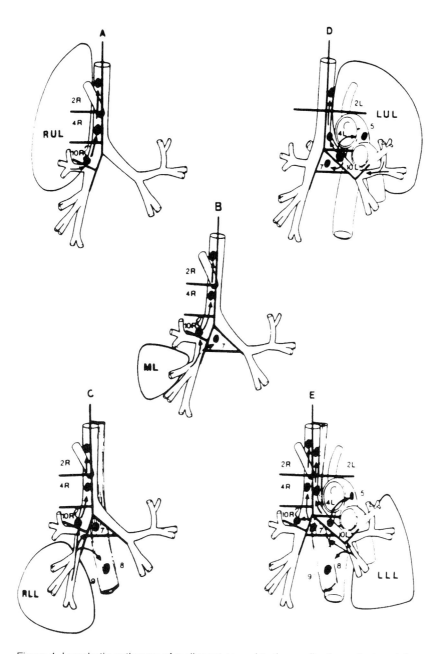

Figure 4. Lymphatic pathways of malignant spread to the mediastinum, by lung lobe of origin.[7]

(b) Left Lung

Left upper lung lobe spreads to 10L, 5, 4L, 2L and 7 (Figure 4d). From region 7 spread occurs to 10R, 4R, 2R. Left lower lung lobe also spreads to 8 and 9. Contralateral involvement via region 7 occurs as frequently in left lung malignancies as ipsilateral spread, in contrast to the right lung, as noted above.

Lymphatic fluid then enters the systemic circulation via the thoracic and right lymphatic ducts, which join the venous system at the root of the left and right neck respectively. The

Table 1. Distribution of Mediastinal and Supraclavicular Lymph Node Metastases by Site of Primary Carcinoma of the Lung[9]

	Percent of Positive Nodes in Region				
Lobe	Subcarinal	Right Paratracheal	Right Supraclavicular	Left Paratracheal	Left Supraclavicular
Right Upper	32	64	30	4	1
Right Lower	46	42	3	12	0
Left Upper	42	16	3	42	39
Left Lower	50	21	4	29	4

locoregional spread of lung cancers can result in specific clinical abnormalities depending on the anatomical structures involved, e.g., external compression of the superior vena cava (SVC) produces SVC syndrome which is a medical emergency. Superior sulcus cancers invading stellate sympathetic ganglion cause Horner's syndrome (Pancoast tumor), while brachial plexus involvement causes ulnar neuropathy. Phrenic or recurrent laryngeal nerve involvement are expressed as diaphragmatic or vocal cord paralysis respectively.

DIAGNOSTIC STUDIES

The following procedures provide essential information for accurate diagnosis and staging of malignant disease. Some of them are also indispensable for RT planning and delivery.

Chest X-Ray — Can identify tumor size and location, effect of bronchial obstruction, diaphragmatic paralysis and invasion of chest wall or mediastinum. Mediastinal lymph nodes > 2 cm in diameter can be detected on plain chest x-ray.[1]

Computed Tomography (CT) — Most accurate radiologic technique for evaluating the mediastinum, mediastinal lymph nodes, pulmonary parenchyma and vertebral bodies. It is not as precise for defining chest wall or rib invasion. Routine concomitant imaging of upper abdomen can delineate metastasis in the liver or adrenals. CT with intravenous infusion of contrast material can highlight invasion or compression of vascular structures. It is highly accurate in predicting likelihood of curative surgical resection in the majority of lung cancer patients. CT also provides essential information regarding the spatial relationship of the carcinoma and regional draining lymph nodes, to the radiosensitive normal structures (e.g., spinal cord, heart) and is presently indispensable in diagnosis, staging, and RT planning. The predictive accuracy of CT in determining mediastinal lymph node metastases is related to the size of the nodes. Lymph nodes <1 cm in greatest diameter on CT are considered to be uninvolved for purposes of clinical staging.[21] Lymph nodes >1 cm and those >1.5 cm in size are involved in 60% to 78%[22,23] and 85% of cases[20] respectively. CT cannot differentiate between malignant, inflammatory, and granulomatous lymphadenopathy or detect microscopic disease in small nodes. Confluence of the cancer with mediastinal structures and thickened mainstem bronchi indicate malignant mediastinal involvement.[24] High resolution CT with 1 mm thick cuts through solitary pulmonary nodules permits precise delineation of anatomic characteristics and identifies presence of calcium which in some cases suggests a benign lesion. CT directed percutaneous needle biopsies of adrenal masses or other lesions may be useful in confirming suspected metastases.

Magnetic Resonance Imaging (MRI) — Provides images in both sagittal and coronal planes and does not expose patients to ionizing irradiation in contrast to CT. Coronal sections can be advantageous in evaluating the aortopulmonary window, a weak area for CT, and involvement of the brachial plexus by superior sulcus carcinomas. MRI offers no advantage over CT in defining mediastinal carcinomatous lymph node involvement.[25] Disadvantages of MRI are poor delineation of small calcifications and prolonged imaging time.

Barium Esophagogram — Indicated for dysphagia to rule out esophageal invasion or a second primary esophageal carcinoma.

Sputum Cytology — This is positive in 60% to 90% of patients with lung cancer who have an abnormal chest x-ray.[26] Higher yields are obtained in patients with central lesions invading major bronchi.

Pleural Effusion Cytology — A pleural effusion found to be malignant can alter staging and treatment. Specimen is obtained by thoracentesis and is usually accompanied by pleural biopsy.

Bronchoscopy — The flexible fiberoptic bronchoscope permits evaluation of segmental and subsegmental bronchi. Examination of the entire tracheobronchial tree excludes a second unsuspected primary cancer and familiarizes the surgeon with any anatomic variations. Combining biopsy and brushings gives an accuracy of 85%.[27] Diagnostic accuracy has not been increased by bronchial washings or postbronchoscopic sputum collection. A 68% positive result has been achieved using brushings and transbronchial biopsy in peripheral lesions >2 cm in diameter.[28] Transthoracic needle aspiration is recommended for lesions <2 cm in size.

Transthoracic Needle Aspiration (TNA) — Useful in diagnosing peripheral pulmonary lesions. A diagnostic accuracy of 88% has been achieved in patients with subsequently proven lung carcinoma depending on the expertise of the physician performing the biopsy and the pathological site.[29] TNA is useful in the diagnosis of superior sulcus carcinomas prior to RT, bilateral lesions and patients who are not surgical candidates.

Bone Marrow Aspiration and Biopsy — Indicated in SCLC since 11% to 40% of patients will have bone marrow involvement.[30]

Pulmonary Function Tests (PFTs) — Important indicators of the patient's ability to undergo surgical resection or withstand RT. A 50% reduction in predicted forced vital capacity (FEV-1), maximum voluntary ventilation (MVV), or vital capacity (VC), necessitates more extensive testing. Arterial blood gases of PaO2 under 65 torr and a PaCO2 over 45 torr are predictors of increased risk. Ventilation and perfusion radionuclide scanning of the lungs (V/Q) are beneficial in assessing operability by measuring the contribution of the lung to be removed to the patient's respiratory ability. An FEV-1 of at least 800 cc is necessary to permit satisfactory rehabilitation.[31] Most patients who can climb two flights of stairs without stopping will tolerate pulmonary resection.

Mediastinoscopy — Essential staging procedure in which mediastinal lymph nodes are evaluated and biopsied. Lymph nodes accessible for biopsy by mediastinoscopy include bilateral tracheal, anterior subcarinal, right and left tracheobronchial. Inaccessible lymph nodes include those located in the anterior mediastinum, aortopulmonary space, posterior carina and posterior/inferior tracheobronchial areas[1] (see Figures 5a and 5b).

Mediastinotomy — Used to biopsy lymph nodes in left aortopulmonary space in the presence of left upper lobe carcinomas. A transverse incision extending from the lateral edge of sternum is placed over the second costal cartilage.

THREE DIMENSIONAL CONFORMAL RADIATION THERAPY FOR LUNG CARCINOMAS

The volume to be treated and irradiation portal configuration are determined by primary malignancy characteristics, lymphatic drainage pattern, histology, available equipment, and the discretion of the attending radiation oncologist. All sites of known carcinomatous involvement and the anatomic regions most likely to be involved by subclinical metastases should be irradiated. Analyses of RTOG studies have provided the most useful information about the extent of the various volumes that need to be irradiated to a specified dose and dose-time pattern.[32] Concurrent chemoradiation is used for SCLC as previously discussed. Irradiation volumes therefore tend to vary slightly from NSCLC since the chemotherapy is expected to eliminate areas of microscopic disease depending on the prevailing malignant cell load.

IMMOBILIZATION/PATIENT POSITION

CT scan is performed with the patient immobilized by Alpha Cradle mold in the supine position and arms folded behind the head to allow use of any oblique treatment angle. Immobilization is essential for daily accurate repositioning and precision of RT delivery.

Localization/Target Delineation

The scanned region includes sufficient margins around area of interest since non-coplanar fields, if utilized, may enter or exit the patient well above or below the anatomical level of the primary target. The supraclavicular areas and diaphragms bilaterally are usually included but extent of region scanned can clearly vary depending on disease location. CT interslice spacing depends on size, shape, location of region of interest, and treatment technique used. The interslice spacing necessary for high resolution 3D reconstruction is generally 3 to 5 mm (3 mm of interslice spacing produces better quality reconstruction). The thickness of each slice is usually 5 mm. The relevant treatment volumes and organs at risk are then outlined on the individual CT slices with a contouring tool. Lung and mediastinal window settings are used to determine gross disease in lung parenchyma and in the mediastinum respectively.

During RT planning and delivery, different volumes may be defined associated with varying concentrations of demonstrated or suspected malignant cell load.

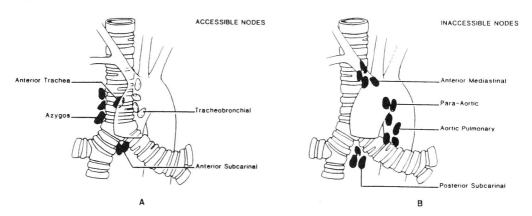

Figure 5. Mediastinal lymph nodes accessible (a) and inaccessible (b) by mediastinoscopy.[21]

(a) Volume Definitions

Two volumes are identified initially: a) gross tumor volume (GTV), and b) organs at risk. Further volumes are subsequently described: c) clinical target volume (CTV), d) planning target volume (PTV).[33] The treated volume (TV) and the irradiated volume (IV), also described in ICRU 50, are not usually defined. They cannot actually be determined until a treatment plan is generated.

GTV — This is all known clinically demonstrable gross malignant disease consisting of primary carcinoma and locoregional disease extent. Regional lymph nodes ≥ 1 cm are included.[34] No GTV can be defined if the carcinoma has been removed prior to radiotherapy. GTV is then the CTV.

CTV — Different CTVs have to be defined for irradiation delivery depending on malignant cell load and treatment policy[33]:

CTV1 — Tissue volume that contains the demonstrable GTV and the surrounding subclinical malignant disease extension which has to be eliminated. CTV is the tissue volume to be irradiated to a specified dose and dose-time pattern to achieve the aim of therapy (cure vs. palliation). A margin of ≤ 1 cm is allowed around GTV for NSCLC to obtain CTV.[34] Chemotherapy is expected to eliminate this subclinical disease in SCLC and this margin around the GTV may therefore not be necessary.

CTV2 — Consists of regional lymph nodes presumed to be involved by subclinical metastatic disease which need to be irradiated electively to a specified dose and dose-time pattern. The following groups of nodes should be included for all primary sites[32,35]:

(1) ipsilateral hilar lymph nodes.
(2) superior mediastinal lymph nodes (i.e., above carina).
(3) subcarinal lymph nodes (with inclusion of contralateral mainstem bronchus).

Additions Depending on Location of Primary Carcinoma

Supraclavicular lymph nodes — Included in upper lobe and mainstem bronchial primaries as well as in the presence of high mediastinal lymphadenopathy. It is acceptable to treat the ipsilateral side only.[15,35]

Inferior mediastinal lymph nodes — Lower lobe carcinomas or inferior mediastinal involvement.

Contralateral hilar lymph nodes — Contralateral mediastinal, subcarinal, and contralateral hilar involvement.

Additional Comments for SCLC

The RTOG includes ipsilateral supraclavicular lymph nodes in SCLC for upper lobe lesions and for bulky (>5 cm) pre- or paratracheal lymphadenopathy detected on contrast enhanced CT only. Contralateral supraclavicular treatment is not allowed. Contralateral hilar RT is allowed only for demonstrable bulky contralateral mediastinal involvement. Mediastinal lymph nodes detected by CT to be ≥ 1.5 cm are also included.[36]

Most clinical groups have been irradiating a large standard prophylactic intrathoracic volume in SCLC even though the chemotherapy usually given is expected to sterilize subclinical disease. This is due to the high incidence of intrathoracic failures observed in most series. Investigators are now reporting that their intrathoracic recurrences are within the RT field suggesting inadequate dose rather than inadequate volume. There is also no difference in intrathoracic control in patients who receive RT, directed to pre- vs. postchemotherapy volumes of disease. The case can presently be made for both large prophylactic intrathoracic and smaller targeted volumes until large trials are designed to detect any subtle differences that may exist between the two approaches.[13] Most patients, however, receive their concurrent RT early in the course of their treatment, before the chemotherapy can cause any significant decrease in the volume of the malignant disease.

PTV — This is an appropriate margin around the CTV and GTV for reproducibility factors (daily repositioning, cardiac and respiratory motion). It is a geometrically defined concept to ensure that the prescribed dose is actually absorbed in the CTV. Margin should be applied conformally in all directions since motion occurs in all three dimensions. Various PTVs can be defined as per above CTVs. PTV is related to the beams through a fixed coordinate system and is used for dose planning and specification. Separate CTVs (or PTVs) may be combined for RT if these volumes are planned to receive an identical dose prescription.

The average treatment variation in chest irradiated areas is 7 to 10 mm.[34] Average movement for upper lung lobe lesions is 2 to 6 mm depending on the presence of chest wall attachment.[37] Hilar, mediastinal, and lower lobe lesions experience more motion due to closer proximity to the heart and diaphragm with average movement of 9.2, 8.7 and 10.1 mm respectively.[37] PTV1 and PTV2 margins are therefore 7 to 10 mm depending on location of the carcinoma. These margins are approximations and may have to be adjusted based on the clinical situation, e.g., if a patient has compromised lung function the size of the different volumes may have to be decreased.

The definitions of TV and IV are provided for completeness.[33]

TV — This is the volume enclosed by a selected isodose surface deemed appropriate to achieve purpose of treatment (e.g., carcinoma eradication or palliation).

IV — This is the tissue volume which receives a dose that is considered significant in relation to normal tissue tolerance (TV + margin for treatment unit penumbra).

(b) Critical Structures (Organs at Risk)

These are normal tissues whose radiation sensitivity may significantly influence treatment planning and/or prescribed dose[33]:

Class I Organs — Radiation lesions are fatal or result in severe morbidity.

Class II Organs — Radiation lesions result in moderate to mild morbidity.

Class III Organs — Radiation lesions are mild, transient, and reversible or result in non-significant morbidity.

Beam Design/Optimization and Virtual Simulation

The isocenter of each treatment group of CTVs is determined at the completion of the localization process and marked on the patient using the laser alignment system. These marks are used to reposition the patient for treatment and serve as a link between virtual simulation and the physical world.[39,40] A series of integrated computerized tools now perform the task of virtual simulation. These tools accurately display the 3D relationships between the target and relevant structures while fully duplicating the functions of a simulator. The target and normal tissue structures required for treatment planning are reconstructed in 3D as wire frame or solid objects and are overlaid on digitally reconstructed radiographs (DRRs), which are computerized pictures produced by passing x-rays through a 3D representation of the CT data using the beam's eye view technique (BEV). Using the structure outlines and the DRRs the treatment fields can be defined. Apertures for the initial anterior and posterior fields are designed using a combination of the DRR radiographic anatomy and the structure outlines.

Table 2. Tolerance of Organs at Risk to Irradiation[15]

Organ at risk	TD 5/5 Volume 1/3	2/3	3/3	Selected End Point (Injury)
Class I				
Normal lung	45 Gy	30 Gy	17.5 Gy	Pneumonitis
Spinal Cord	50 Gy/5cm	50 Gy/10 cm	47 Gy/20 cm	Myelitis/necrosis
Heart	60 Gy	45 Gy	40 Gy	Pericarditis
Liver	50 Gy	35 Gy	30 Gy	Hepatitis
Class II				
Brachial plexus	62 Gy	61 Gy	60 Gy	Clinical dysfunction
Esophagus	60 Gy	58 Gy	55 Gy	Stricture/perforation
Class III				
Rib cage	50 Gy	...	...	Pathologic fracture
Skin ...	...	50 Gy/100 cm^2		Telangiectasia
Muscle (adult)	...	...	<10 Gy	Atrophy/fibrosis
Large vessels	...	..	<8 Gy	Atherosclerosis

The structure outlines and DRR information are also used to geometrically optimize the direction, position, and shape of the cone-down fields. All the contour and field information is then sent to the treatment planning system for dose optimization which may require modification of the treatment fields. The virtual simulator is then used to generate DRRs of the final treatment fields. Currently all the lung cancer patients also have a conventional simulation of the final treatment fields because of inadequate DRR quality.

Dose Prescription and Technical Factors

Irradiation Dose and Dose-Time Parameters

(a) *NSCLC* — A total dose of 60 Gy in 2 Gy daily fractions five times a week is delivered to clinically involved areas. Areas with presumed microscopic disease receive approximately 45 Gy with the same daily fraction size.[41] Parallel opposed anteroposterior fields are used initially to encompass PTV2 until the maximum dose to the spinal cord is 45 Gy. The rest of the

Table 3. RTOG Maximum Doses to Critical Structures [35,36]

Organ at risk		Max. Dose NSCLC	Max. Dose SCLC
Normal lung	(ipsilateral)	25 Gy	
	(entire ipsilateral)		10 Gy
Normal lung	(contralateral)	20 Gy	15 Gy
Spinal Cord	(maximum dose)	45 Gy	36 Gy
Heart	(100%)	45 Gy	36 Gy
	(<50%)	50 Gy	

total dose to the clinically involved areas is then delivered with reduced fields to PTV1 with an alternate beam arrangement which avoids the spinal cord. Fraction size is reduced to 1.8 Gy daily when concurrent chemotherapy is also given, for a total dose to PTV2 of 45 Gy. PTV1 then receives a further 18 Gy in 2 Gy daily fractions with reduced fields as before for a cumulative total of 63 Gy. There are no heterogeneity corrections used in the definition of these doses.[35] Dose is prescribed to isodose line which encompasses the PTV adequately to ensure absorption of prescribed dose by CTV while maintaining a dose uniformity of ±5%. This is usually the 95% isodose line or higher.

(b) SCLC — A total dose of 45 Gy at 1.8 Gy daily fractions is commonly delivered. Parallel opposed anterioposterior fields are used initially until the maximum dose to the spinal cord is 36 Gy and then an alternate beam arrangement delivers the rest while avoiding the spinal cord.[41] Hyperfractionated RT regimes are also being utilized with promising results and are currently actively investigated.

Beam Energy — Megavoltage equipment with minimum peak photon energies of 6 MV is used. Electrons with at least 90% dose at 3 cm depth may be used to treat the supraclavicular lymph nodes and should be specified to D_{max}.

Beam Shaping — The field is fashioned using custom 5 HVL blocks which define the irradiated volumes and protect organs at risk. Posterior spinal cord block is not acceptable.

Treatment Distance — Minimum isocenter distance should be 100 cm for SAD techniques. Minimal treatment distance to skin should be greater than 80 cm for SSD techniques.

Compensating Filters and Wedges — These are often used to improve dose uniformity especially in areas with large sloping contours which are usually encountered during the treatment of upper lung lobe carcinomas.

Treatment Plan Verification

DRRs, as previously discussed, are computerized pictures produced by passing x-rays through a 3D representation of the CT data in the BEV mode.[40] Films are produced of the DRRs with the overlaid marks which indicate the specific treatment radiation portal characteristics and selected anatomical contours (target, organs at risk). Anterior and lateral orthogonal set up films are also produced. The use of only the DRRs for treatment verification is unsatisfactory because they do not provide sufficient information regarding critical soft tissue organs. Conventional simulator films depict anatomical structures with better definition than DRRs. The patient therefore undergoes conventional simulation according to the treatment parameters defined by the treatment plan. Radiographs are obtained and compared with the anterior and lateral set up DRRs to confirm correct positioning. The isocenter as predicted on the DRRs and as achieved on the physical x-rays is then compared and appropriate adjustments made.

Radiographs are obtained for each of the treatment fields. The simulator and DRR films are aligned using the isocenter location and bony landmarks. The relevant volumes of interest already outlined on the DRRs are now transferred onto the simulator films, which are used as templates for cerrobend block fabrication. The simulator films are also used for comparison with treatment portal verification films. Conventional simulator films enable more accurate treatment verification (e.g., block positioning and isocenter definition) due to the better anatomical delineation of the different soft and bony tissues within the radiation field.

CASE ILLUSTRATION

A 60-year-old female smoker with cough and intermittent hemoptysis was found to have a left upper lung lobe bronchial lesion and left hilar lymphadenopathy on CT. The mediastinum was clear but there was fluid in the left pleural space. Bronchoscopy was confirmatory and biopsy showed NSCLC (squamous cell carcinoma). Cytology of the left pleural space fluid was negative for malignant cells. Staging investigations did not show any metastases but the patient refused thoracotomy. She was diagnosed as Clinical Stage II (T2N1M0) NSCLC (squamous cell carcinoma) of left upper lung lobe. She agreed to undergo definitive RT.

Treatment Planning — Proceeded according to 3D CRT technique as already described.

Treatment Modality — Definitive 3D CRT. The disease appeared to be localized with no metastases.

Treatment Position — Supine with arms folded above head.

Immobilization — Alpha Cradle body mold.

Localization of Treatment Volumes — Figures 6 through 8 show delineation of volumes on axial CT slices at the level of the demonstrated carcinoma. Figure 9 is a coronal DRR of PTV2, CTV2, CTV1 and GTV. Figure 10 is a coronal DRR of PTV1.

GTV   Primary endobronchial squamous cell carcinoma and left hilum (see Figures 6 and 9).

CTV1   GTV + 5 mm margin for subclinical disease extension (see Figures 6 and 9).

PTV1   CTV1 + 7 mm margin for limitations of treatment technique (see Figure 10).

CTV2   Regional lymph nodes-superior mediastinal, subcarinal and contralateral main stem bronchus, supraclavicular ipsilateral lymph nodes (see Figures 7 and 9).

PTV2   CTV1 + CTV2 + 7 mm margin (see Figures 8 and 9).

Organs at Risk — Class I (spinal cord, heart, ipsilateral/contralateral normal lung), class II (brachial plexus, esophagus), class III (skin, ribs, muscle, large vessels). Tolerance doses

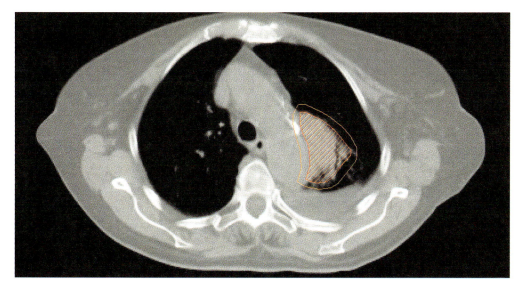

Figure 6. One transverse CT-section is shown at the level of the demonstrated carcinoma which involved the left hilar region (GTV) and is shown by the striated area. A margin was added around the GTV to include presumed local subclinical malignant disease extent (CTV I). GTV: Red. CTV I: Brown.

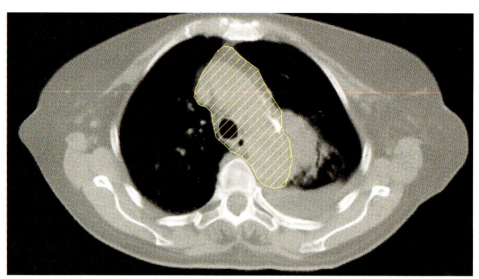

Figure 7. There were no demonstrable clinical mediastinal or supraclavicular lymph node metastases. These areas and the contralateral main bronchus are considered to be at high risk for subclinical malignant disease extent (CTV II) and are electively irradiated. CTV II: Yellow.

were not to be exceeded as already discussed. The spinal cord, represented in green, was the only organ at risk outlined on the DRRs (Figures 9 and 10).

Irradiation Dose

PTV1 — 16 Gy in 8 fractions over ± 2 weeks.

PTV2 — 44 Gy in 22 fractions over ± 4.5 weeks. Dose was prescribed to the isodose line

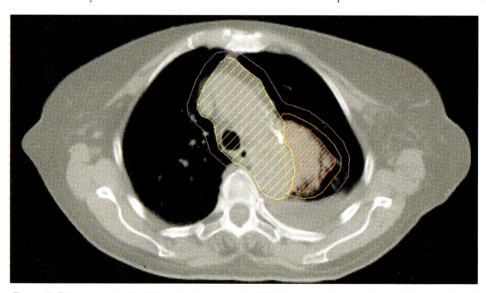

Figure 8. The geometrical relationship between the CTV, the rest of the patient, and the treatment beam(s) should be stable and not change during or between treatment. This relationship should be correlated to a fixed coordinate system related to a point in/at the patient. The CTV will experience daily minor random variations in position. Treatment, therefore, has to be planned for a larger volume than the CTV and a suitable PTV is defined for treatment planning purposes and for dose recording/reporting. GTV: Red. CTV II: Brown. CTV I: Yellow. PTV II: Purple.

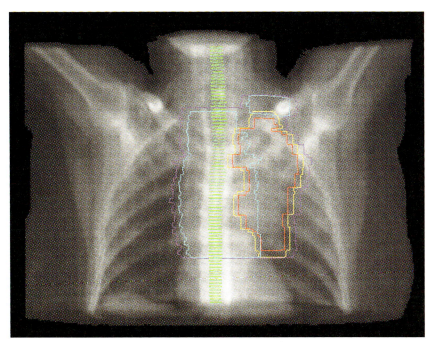

Figure 9. DRR (coronal view). GTV: Red. CTV II: Blue. CTV I: Yellow. PTV II: Purple.

which encompassed the PTV adequately to ensure absorption of prescribed dose by the CTV while maintaining a dose uniformity of ± 5%. This is usually the 95% isodose line or higher.

Technical Factors

Energy utilized — 6 MV photons.

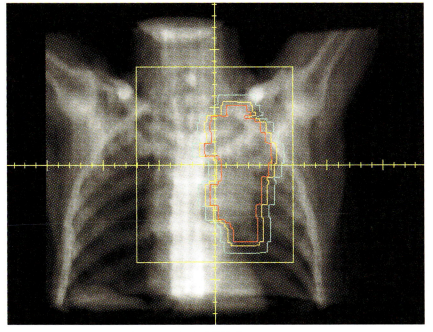

Figure 10. DRR of PTV I (coronal view). PTV I: Blue.

Beam arrangement — PTV2. A pair of opposed equally weighted beams with direction 0° and 180° were used. The technique was isocentric SAD 100 cm. Dose was prescribed to the isodose line which included the PTV2.

PTV1. An oblique beam arrangement was used to keep delivered doses to organs at risk within the previously discussed limits.

Homogeneity — Wedges were used as necessary to improve the distribution of dose uniformity. Lung inhomogeneity correction was not done.

Fashioning of field — 5 HVL cerrobend blocks were used to define the irradiated volume and protect organs at risk.

Control Measures — Treatment portal verification films were taken weekly.

CONCLUSION

Conformal three dimensional RT provides superior delivery of RT with reduced risk to normal tissues. This approach may make RT dose escalation possible with the potential to improve the therapeutic ratio for lung carcinomas.

REFERENCES

1. Coia LR, Moylan DJ. Introduction to Clinical Radiation Oncology, 2nd edition. Madison, Wisconsin: Medical Physics Publishing, 1994.

2. Holleb AI, Fink, DJ, Murphy GP. American Cancer Society Textbook of Clinical Oncology. Atlanta, Georgia: American Cancer Society, 1991.

3. Weisenburger TH, Gail M. The lung cancer study group: effects of postoperative mediastinal radiation on completely resected Stage II and III epidermoid cancer of the lung. N Engl J Med 315:1377-1381;1986.

4. Dillman RO, Seagren SL, Herndon J, et al. Randomized trial of induction chemotherapy plus radiation therapy versus RT alone in stage III non-small cell lung cancer. Five year follow-up of CALGB 84-33. Proc Amer Soc Clin Oncol 13:329; 1993. Abstract.

5. Schaake-Koning C, van der Bogaert W, Dalesio O, et al. Effects of concomitant cisplatin and radiotherapy on inoperable non-small cell lung cancer. N Engl J Med 326:524-530;1992.

6. Le Chevalier T, Arriagada R, Tarayre M, et al. Significant effect of adjuvant chemotherapy on survival in locally advanced non-small cell lung carcinoma. J Nat Canc Inst (Ltr) 84:58; 1992.

7. Murray N, Coy P, Pater JL, et al. Importance of timing for thoracic irradiation in the combined modality treatment of limited-stage small-cell lung cancer. J Clin Oncol 11:336-344; 1993.

8. Pignon JP, Arriagada R, Ihde DC, et al. Meta-analysis of thoracic radiotherapy for small-cell lung cancer. N Engl J Med 327:1618-1624;1992.

9. Einhorn LH. The case against prophylactic cranial irradiation in limited small cell lung cancer Semin Radiat Oncol 5:57-60;1995.

10. Ball DL, Matthews JP. Prophylactic cranial irradiation: more questions than answers. Semin Radiat Oncol 5:61-68;1995.

11. Arriagada R, LeChevalier T, Borie F, et al. Randomized trials on prophylactic cranial irradiation for patients with small cell lung cancer in complete remission. Lung Cancer (suppl 2) 11:177-178; 1994.

12. Perez CA, Stanley K, et al. Impact of irradiation technique and tumor extent in tumor control and survival of patients with unresectable non-oat cell carcinoma of the lung. Cancer 50:1091-1099;1982.

13. Lichter AS, Turrisi III AT. Small cell lung cancer: the influence of dose and treatment volume on outcome. Semin Radiat Oncol 5:44-49;1995.

14. Leibel SA, Kutcher GJ, et al. Three dimensional conformal radiation therapy at the Memorial Sloan-Kettering Cancer Center. Semin Radiat Oncol 2:274-289;1992.

15. Armstrong JG, Burman C, Leibel S, et al. Three-dimensional conformal radiation therapy may improve the therapeutic ratio of high dose radiation therapy for lung cancer. Int J Radiat Oncol Biol Phys 26:685-689;1993.

16. Basmajian JF, Slonecker CE. Grant's Method of Anatomy: A Clinical Problem-Solving Approach, 11th edition. Baltimore, Maryland: Williams & Wilkins, 1991.

17. American Joint Committee on Cancer: Manual For Staging of Cancer, 4th edition. Philadelphia, Pennsylvania: J.B. Lippincott Company, 1992.

18. American Thoracic Society clinical staging of primary lung cancer. Am Rev Resp Dis 127:659-664;1983.

19. DeVita Jr VT, Hellman S, Rosenberg SA. Cancer: Principles & Practice of Oncology, 4th edition. Vol. 1. Philadelphia, Pennsylvania: J.B. Lippincott, 1993.

20. Buy NJ, Ghossain MA, et al. Computed tomography of mediastinal lymph nodes in non-small cell lung cancer J Comput Assist Tomogr 12:545-552;1988.

21. Cox JD. Moss' Radiation Oncology: Rationale, Technique, Results, 7th edition. St. Louis, Missouri: Mosby, 1994.

22. Sagel SS. Lung, pleura, pericardium and chest wall. In: Lee JKT, Sagel SS, Stanley, RJ, eds., Computed Body Tomography. New York: Raven Press, 1993.

23. McKenna Jr RJ, Libshitz HI, Mountain CE, et al. Roentgenographic evaluation of mediastinal nodes for preoperative assessment in lung cancer. Chest 88:206-210;1985.

24. Glazer GM, Orringer MB, Gross BH, et al. The mediastinum in non-small cell lung cancer: CT surgical correlation. Am J Roentgenol Radium Ther Nucl Med 142:1101-1105; 1984.

25. Daly BDT, Faling LJ. Pugatch RD, et al. Computed tomography: an effective technique for mediastinal staging in lung cancer. J Thorac Cardio Surg 88:486-494;1984.

26. Frederick HM, Bernardino ME, Baron M, et al. Accuracy of chest computerized tomography in detecting malignant hilar and mediastinal involvement by squamous cell carcinoma of the lung. Cancer 54:2390-2395;1984.

27. Heelan RT, Martini N, Westcott JW, et al. Carcinomatous involvement of the hilum and mediastinum: computed tomographic and magnetic resonance evaluation. Radiology 156: 111-115; 1985.

28. Oswald NC, Hinson KFW, Canti G, et al. The diagnosis of primary lung cancer with special reference to sputum cytology. Thorax 16:623-631;1971.

29. Zavala DC. Diagnostic fiberoptic bronchoscopy: technique and results of biopsy in 600 patients. Chest 68:12-19;1975.

30. Radke JR, Conway WA, Eyler WR, et al. Diagnostic accuracy in peripheral lung lesions: factors predicting success with flexible fiberoptic bronchoscopy. Chest 76:176-179; 1979.

31. Zaman MB, Hajder S, Melamed MR, et al. Transthoracic aspiration cytology of pulmonary lesions. Sem Diag Pathol 3:176-187;1986.

32. Ihde DC, Hansen HH. Staging procedures and prognostic factors in small cell carcinoma of the lung. In: Greco FA, Oldham RK, Bunn PA, eds. Small Cell Lung Cancer. New York: Grune and Stratton, 1981, pp 261-286.

33. Olsen GN, Block J, Tobias JA. Prediction of post pneumonectomy function using quantitative macroaggregate lung scanning. Chest 66:13-16;1974.

34. Perez CA, Brady LW. Principles and Practice of Radiation Oncology, 2nd edition. Philadelphia, Pennsylvania: J.B. Lippincott, 1992.

35. ICRU Report 50. Prescribing, Recording and Reporting Photon Beam Therapy. Bethesda, Maryland: International Commission on Radiation Units and Measurements, 1993.

36. Graham MV, Matthews JW, Harms Sr WB, et al. Three-dimensional radiation treatment planning study for patients with carcinoma of the lung. Int J Radiat Oncol Biol Phys 29:1105-1117;1994.

37. Curran Jr WJ, Langer CJ, et al. Radiation Therapy Oncology Group RTOG 94-10: a three-arm phase III study of concomitant versus sequential chemotherapy and thoracic radiotherapy for patients with locally advanced inoperable non-small cell lung cancer. 1994.

38. Glisson BS, Komaki R. Radiation Therapy Oncology Group RTOG 93-12. Phase II study of cisplatin, ifosfamide/mesna, etoposide, and concurrent accelerated hyperfractionated thoracic radiotherapy for patients with limited small cell lung cancer. 1994.

39. Ross CS, Hussey DH, Pennington EC, et al. Analysis of movement of intrathoracic neoplasms using ultrafast computerized tomography. Int J Radiat Oncol Biol Phys 18: 671-677; 1990.

40. Emami B, Lyman J, Brown A, et al. Tolerance of normal tissue to therapeutic irradiation. Int J Radiat Oncol Biol Phys 21:109-122;1991.

41. Lichter AS, Sandler HM, et al. Clinical experience with three-dimensional treatment planning. Semin Radiat Oncol 2:257-266;1992.

42. Sailer SL, Chaney, EL, et al. Treatment planning at the University of North Carolina at Chapel Hill. Semin Radiat Oncol 2:274-289;1992.

Chapter 11

CERVICAL & ENDOMETRIAL CANCERS

Rachelle M. Lanciano, MD and Steven Bonin, MD

INTRODUCTION

Cervix and uterine cancer comprise 62% of all female genital malignancies with an estimated 44,500 new cases in the United States in 1993.[1] The treatment of cervical carcinoma involves external irradiation (RT) prior to brachytherapy in almost all cases with significant bulk of disease (e.g., stage IB and greater). The goal of external RT is to sterilize microscopic deposits in the pelvic ± paraaortic lymph nodes (LN) and reduce the bulk of central disease prior to brachytherapy. The treatment of endometrial cancer with pelvic RT is typically following total abdominal hysterectomy (TAH) and bilateral salpingoophorectomy (BSO) to reduce the risk of nodal and vaginal recurrence which ranges from 20% to 30% in high risk patients.[2,3]

For both cervical and endometrial cancer, a local recurrence is rarely salvaged following RT; therefore, optimization of RT to the target should be the goal of CT scan treatment planning.[4,5] Typically, pelvic RT portals are designed by using bony landmarks as indirect markers of pelvic LN location and local disease extension. For example, the Gynecologic Oncology Group (GOG) has recommended a 1 to 1.5 cm margin on the pelvic rim for anteroposterior fields, and a margin defined by a vertical line from the anterior aspect of the pubic symphysis on lateral fields.[6] Other investigators have brought this practice into question, suggesting that some pelvic failures may be due to inadequate coverage of pelvic nodal regions. Greer et al. performed intraoperative measurements on 100 patients with gynecologic malignancies. They concluded that standard RT technique may have been insufficient vis-a-vis LN at risk in as many as one-half of their patients.[7] Pendlebury et al. reviewed the simulation films of patients treated with stage II and III cervical cancer. In those patients who had bipedal lymphangiogram, modifications of the "standard" portals were required in 62% in order to ensure adequate nodal coverage.[8] Bonin et al. confirmed the inadequacy of bony landmarks as indirect markers of LN location. Forty-five percent of patients in this study had LN inadequately covered with standard GOG fields.[9]* Therefore, CT simulation allows individualization of treatment fields by providing the radiotherapist with the anatomic information needed to design fields and blocks that will insure coverage of LN and local disease extension at risk while minimizing dose to normal tissues.

---

*Kim et al. found 62% of patients had inadequate margin on the posterior border (52-53 interspace on the rectal block) or anterior border (due to an enlarged uterus) on lateral pelvic portals when CT simulations of the pelvis were retrospectively applied. A higher pelvic failure rate was present for those patients with an inadequate margin.[10] Russell et al., using sagittal MRI, found inadequate margin on lateral pelvic portals in 56% as well as inadequate coverage of the uterine fundus in 62% of patients if standard fields were used.

ANATOMY AND PATTERNS OF SPREAD

The pelvic fields for external RT are defined by the lymphatics that drain the cervix and uterus. The draining lymphatics can be divided into two major groups: external iliac and internal iliac (hypogastric) systems. Mangan et al.[12] sent 175 diagrams to academic pathologists, radiation oncologists and gynecologic oncologists and requested names of each nodal group listed in Figure 1. The most inconsistently labeled LN group was #6, where responses included gluteal, hypogastric and obturator, respectively. Mangan proposed a clinical nomenclature of the pelvic LN groups based on the artery dissected to obtain them.[12] Cherry et al.[13] performed radical hysterectomy and bilateral pelvic lymphadenectomy on 213 patients with stage I and II squamous cell carcinomas of the cervix. They found 43% to have pelvic LN metastasis. He described the distribution of the pathologically proven metastatic LN by site in the pelvis. The external iliac (described as medial, anterior and lateral groups) corresponded to 47% of the total number of involved nodal groups, while the obturator and hypogastric nodes were less common with 20% and 7%, respectively, of the total number of involved nodal groups.[13] Twiggs et al. described a 34% risk of pathologically positive LN from pelvic lymphadenectomies prior to definitive RT for stages IB through IIIB cervix cancer.[14] From the GOG experience, Heller et al. described a 24% risk of involvement of the common iliac/paraaortic LN for stages IIB through IVA cervix cancer after selective paraaortic and common iliac lymphadenectomy prior to RT.[15] For clinical stage I and II endometrial cancer, the risk of pelvic and paraaortic LN are dependent on the grade and depth of myometrial invasion ranging from 3% and 1% with <1/3 invasion and Grade I, to 34% and 23% for >2/3 invasion and Grade III, respectively.[16]

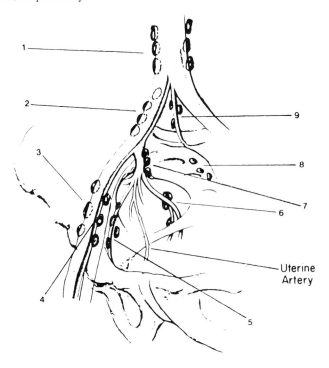

Figure 1. 1 indicates paraaortic; 2, common iliac; 3, external iliac, lateral group; 4, external iliac, medial group; 5, obturator; 6, inferior gluteal; 7, hypogastric; 8, presacral; and 9, subaortic.

Therefore, there is a significant risk of pelvic and paraaortic LN involvement in cervix and endometrial cancer. Defining LN location by CT simulation and the use of conformal blocks insures adequate coverage of LN at risk and minimizes the volume of normal tissues irradiated.

DIAGNOSTIC STUDIES

Physical examination remains the most important pretreatment evaluation in defining the local extent of the cervical cancer into the parametrium and vagina and is the only gynecologic malignancy that remains clinically staged. Cystoscopy and proctoscopy have low yield and are only recommended for stages IIB and greater cervix cancer. CT scan of the abdomen and pelvis has replaced intravenous pyelography (IVP) for evaluating the upper urinary tracts and can also diagnose bulky adenopathy or local disease. Lymphangiography (LAG) is the most accurate non-surgical means of evaluating the paraaortic and common iliac region and if negative the risk of positive LN is less than 8%.[15] Extraperitoneal selective pelvic and paraaortic lymphadenectomy remains the gold standard for evaluating LN in cervix cancer (Figure 2).

The GOG trials require paraaortic and common iliac lymphadenectomy prior to RT, since involvement of these nodes is the most important negative prognostic factor for cervical cancer.[17] The FIGO staging system for cervical cancer allows physical examination, cystoscopy, proctoscopy, IVP, routine blood studies, and chest x-ray (Table 1).

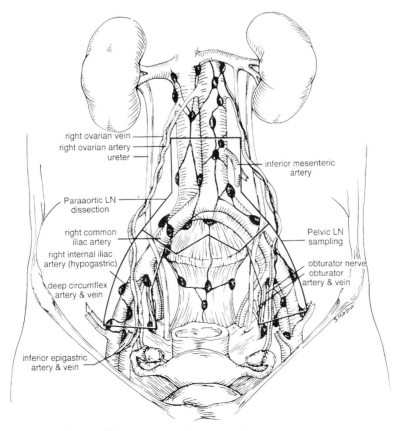

Figure 2. Anatomic boundaries of paraaortic/common iliac and pelvic lymphadenectomies.

Table 1. Definition of TNM — Cervical Cancer

Primary Tumor (T)

TNM	FIGO	Definition
TX	...	Primary tumor cannot be assessed.
T0	...	No evidence of primary tumor.
Tis	...	Carcinoma in situ.
T1	I	Cervical carcinoma confined to the uterus (extension to the corpus should be disregarded).
T1a	IA	Preclinical invasive carcinoma, diagnosed by microscopy only.
T1a1	IA1	Minimal microscopic stromal invasion.
T1a2	IA2	Tumor with an invasive component 5 mm or less in depth taken from the base of the epithelium and 7 mm or less in horizontal spread.
T1b	IB	Tumor larger than T1a2.
T2	II	Cervical carcinoma invades beyond the uterus but not to the pelvic wall or to the lower third of the vagina.
T2a	IIA	Tumor without parametrial invasion.
T2b	IIB	Tumor with parametrial invasion.
T3	III	Cervical carcinoma extends to the pelvic wall and/or involves the lower third of the vagina and/or causes hydronephrosis or nonfunctioning kidney.
T3a	IIIA	Tumor involves the lower third of the vagina, with no extension to the pelvic wall.
T3b	IIIB	Tumor extends to the pelvic wall and/or causes hydronephrosis or a nonfunctioning kidney.
T4*	IVA	Tumor invades the mucosa of the bladder or rectum and/or extends beyond the true pelvis.
M1	IVB	Distant metastasis.

*Note: Presence of bullous edema is not sufficient evidence to classify a tumor as T4.

Regional Lymph Nodes (N)

NX		Regional lymph nodes cannot be assessed.
N0		No regional lymph node metastasis.
N1		Regional lymph node metastasis.

Distant Metastasis (M)

TNM	FIGO	Definition
MX		Presence of distant metastasis cannot be assessed.
M0		No distant metastasis.
M1	IVB	Distant metastasis.

pTNM Pathologic Classification
The pT, pN, and pM categories correspond to the T, N, and M categories.

Table 1 (continued). Stage Grouping Carcinoma of the Cervix

AJCC/UICC				FIGO
Stage 0	Tis	N0	M0	
Stage IA	T1a	N0	M0	Stage IA
Stage IB	T1b	N0	M0	Stage IB
Stage IIA	T2a	N0	M0	Stage IIA
Stage IIB	T2b	N0	M0	Stage IIB
Stage IIIA	T3a	N0	M0	Stage IIIA
Stage IIIB	T1	N1	M0	Stage IIIB
	T2	N1	M0	
	T3a	N1	M0	
	T3b	Any N	M0	
Stage IVA	T4	Any N	M0	Stage IVA
Stage IVB	Any T	Any N	M1	Stage IVB

Endometrial cancer is surgically staged and prognosis is dependent not only on pathologic stage (e.g., extrauterine involvement, pelvic and paraaortic LN involvement, and deep invasion of the cervix and myometrium) but also on grade, histology, and capillary lymphatic space invasion. Pelvic control, distant control, and survival following total abdominal hysterectomy decreases as the number of poor prognostic factors increases.[18,19] Vaginal failure is dependent as well on depth of invasion and grade and for high risk patients (myometrial invasion greater than one-third, grade 3 or cervical involvement) is 16% to 20% at 10 years.[3] The FIGO staging system for endometrial cancer is shown in Table 2.

IMMOBILIZATION

Patients undergo CT simulation in the prone or supine position with their arms overhead. For immobilization, a custom Alpha Cradle from the iliac crest to the knees is made. Patients are given 450 cc of ready-to-use barium sulfate (1.2%) suspension for computed tomography (Readi-CAT, E-Z-M Inc). The patient is instructed to drink the entire bottle one hour before casting so that contrast is optimized in the small bowel at the time of simulation (approximately one and a half hours following ingestion). A large Q-tip is soaked in Readi-CAT, placed in a glove, and inserted into the vagina at the apex for endometrial cancer patients and at the inferior aspect of the cervix for cervical cancer patients.

Each patient is scanned from the top of the iliac crest to 1 cm below the ischial tuberosity. The patient is imaged with overlapping CT slices that are 5 mm thick at 3 mm intervals.

Table 2. Definition of TNM — Endometrial Cancer

Primary Tumor (T)

TNM	FIGO	Definition
TX	...	Primary tumor cannot be assessed.
T0	...	No evidence of primary tumor.
Tis	...	Carcinoma in situ.
T1	I	Tumor confined to the corpus uteri.
T1a	IA	Tumor limited to the endometrium.
T1b	IB	Tumor invades up to or less than one-half of the myometrium.
T1c	IC	Tumor invades more than one-half of the myometrium.
T2	II	Tumor invades the cervix but not extending beyond the uterus.
T2a	IIA	Endocervical glandular involvement only.
T2b	IIB	Cervical stromal invasion.
T3	III	Local and/or regional spread as specified in and/or N1T3a, b, N1 and FIGO IIIA, B, and C below.
T3a	IIIA	Tumor invades the serosa and/or adnexa (direct extension or metastasis) and/or cancer cells in ascites or peritoneal washings.
T3b	IIIB	Vaginal involvement (direct extension or metastasis).
N1	IIIC	Metastasis to the pelvic and/or paraaortic lymph nodes.
T4*	IVA	Tumor invades the bladder mucosa or the rectum and/or the bowel mucosa.
M1	IVB	Distant metastasis (excluding metastasis to the vagina, pelvic serosa, or adnexa; including metastasis to intraabdominal lymph nodes other than paraaortic, and/or inguinal lymph nodes).

*Note: Presence of bullous edema is not sufficient evidence to classify a tumor as T4.

Regional Lymph Nodes (N)

NX		Regional lymph nodes cannot be assessed.
N0		No regional lymph node metastasis.
N1		Regional lymph node metastasis.

Distant Metastasis (M)

TNM	FIGO	Definition
MX		Presence of distant metastasis cannot be assessed.
M0		No distant metastasis.
M1	IVB	Distant metastasis.

pTNM Pathologic Classification
The pT, pN, and pM categories correspond to the T, N, and M categories.

CERVICAL & ENDOMETRIAL CANCERS

Table 2 (continued). Stage Grouping Carcinoma of the Endometrium

AJCC				FIGO
Stage 0	Tis	N0	M0	
Stage IA	T1a	N0	M0	Stage IA
Stage IB	T1b	N0	M0	Stage IB
Stage IC	T1c	N0	M0	Stage IC
Stage IIA	T2a	N0	M0	Stage IIA
Stage IIB	T2b	N0	M0	Stage IIB
Stage IIIA	T3a	N0	M0	Stage IIIA
Stage IIIB	T3b	N0	M0	Stage IIIB
Stage IIIC	T1	N1	M0	Stage IIIC
	T2	N1	M0	
	T3a	N1	M0	
	T3b	N1	M0	
Stage IVA	T4	Any N	M0	Stage IVA
Stage IVB	Any T	Any N	M1	Stage IVB

LOCALIZATION

Definition of Gross Tumor Volume, Clinical Target Volume, and Planning Target Volume

The GTV1 represents the cervix/uterus/parametrium and vaginal extension of the cervical cancer. There is no GTV1 for endometrial cancer following TAH/BSO. The CTV1 includes the GTV1 with a 5 cm margin of uninvolved vagina and the internal/external ± common iliac LN for both cervix and endometrial cancer.

The planning target volume (PTV1) represents the CTV1 plus a margin necessary for day-to-day set up variability. At Fox Chase Cancer Center the daily set up variability is 1 cm for the pelvis which represents two standard deviations around the mean.[20]

The PTV1 is typically treated with a four-field plan with shaped blocks to assure that at least the 95% isodose covers the PTV1. The dose to the PTV1 is 4000 to 5000 cGy which represents the minimum tumor dose.

DEFINITION OF CRITICAL STRUCTURES

The critical structures of concern in the pelvis include the bladder, rectum, femoral head, and small bowel which have maximum tolerable doses of 6500 cGy, 6000 cGy, 5200 cGy, and 5000 cGy, respectively.[21] Since typical doses delivered to the pelvis for cervical and endometrial cancer are in the range of 4000 to 5000 cGy, no critical structure is above the TD5/5 for external RT; but the rectum, bladder, and small bowel reach tolerance following brachytherapy. Patients treated with RT following hysterectomy tend to have more small bowel in the treatment field since the uterus no longer is present to displace small bowel superiorly and anteriorly. Therefore, when patients are treated with RT following hysterectomy, small bowel studies are necessary at simulation to evaluate the volume of small bowel in the treatment field. Techniques to exclude small bowel such as anterior compression, bladder distension or superior field reduction should be used to minimize late bowel complications.

TREATMENT VERIFICATION

Digitally reconstructed radiographs (DRRs) alone provide enough information for treatment verification, since pelvic bones such as the pubic symphysis, iliac crest, acetabulum and sacroiliac joint are well delineated on a DRR. Contrast in small intestine and the vaginal marker are not seen well on the DRR; therefore, if indicated, these markers need to be contoured for visualization on DRR. Normal structures such as kidney, rectum, bladder or small bowel should be contoured for visualization on DRR.

CASE ILLUSTRATIONS

Case #1

JB is a 52-year-old G4 P4 postmenopausal white female who presented with abnormal postmenopausal bleeding while on estrogen replacement therapy, dyspareunia, and pelvic pain. Hysteroscopy revealed abnormal-appearing tissue from either the lower uterine segment or the upper endocervix and curettage revealed adenocarcinoma. The patient underwent radical hysterectomy, BSO with bilateral pelvic, common iliac, and lower paraaortic lymphadenectomy. During lymphadenectomy no suspicious LN were encountered grossly; however, pathology revealed bilateral pelvic and unilateral paraaortic LN metastases. The moderately differentiated adenocarcinoma was arising from the endocervix with penetration of the entire thickness of the cervical wall, although the cervical size was only 2.2 cm. The patient was treated with 3 cycles of Adriamycin/Cisplatinum, followed by pelvic and paraaortic RT (4500 cGy) and a brachytherapy vaginal boost (Figures 3 and 4). In Figure 3, note the location of the LN dissection by vascular clips placed around the major vessels.

Case #2

MW is an 85-year-old white female who presented with postmenopausal vaginal bleeding. Fractional dilation and curettage revealed Grade II adenocarcinoma consistent with an endometrial primary. Physical examination confirmed an erythematous lesion in the cervical os with a normal sized uterus. The patient was treated with preoperative pelvic RT (4400 cGy), followed by one intracavitary insertion with a Fletcher-Suit applicator (2500 cGy to Point A) (Figures 5 and 6). Figure 5 shows the GTV1, which included the uterus and cervix. CTV1 included the GTV1 with 5 cm margin plus internal and external iliac nodal regions.

CERVICAL & ENDOMETRIAL CANCERS

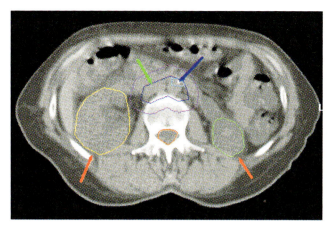

Figure 3a. Cervix cancer following radical hysterectomy with positive paraaortic LN. Note kidneys (red arrow), inferior vena cava (green arrow), and aorta (blue arrow). Due to the anterior position of the kidney, patient was treated with anterior and posterior opposed fields. Note surgical clips on vasculature.

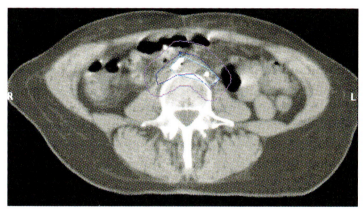

Figure 3b. Common iliac veins/arteries with clips along vasculature corresponding to region of LN dissection.

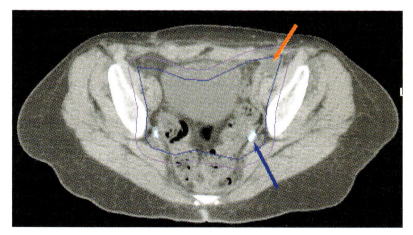

Figure 3c. CTV covers anterior external iliac vessels with lymphocyst present (red arrow) following dissection. Note posterior location of clips surrounding internal iliac (hypogastric) vessels (blue arrow).

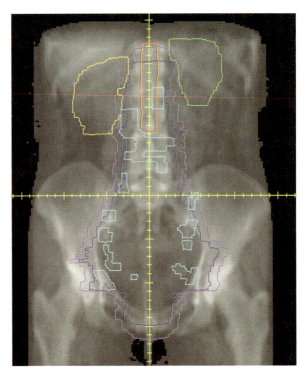

Figure 4a. Pelvis and paraaortics with clips on vasculature (light blue), right kidney (yellow), left kidney (green), spinal cord (red). Note lateral extent of external iliac vessels. CTV1: Dark blue. PTV1: Purple.

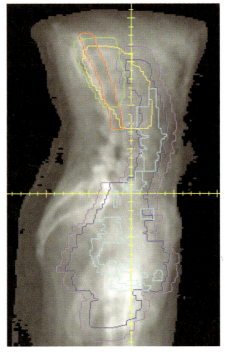

Figure 4b. Lateral DRR. Note anterior location of kidney.

CERVICAL & ENDOMETRIAL CANCERS

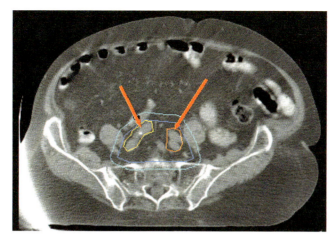

Figure 5a. Stage II adenocarcinoma of the uterus with gross cervical extension. Note common iliac vessels defining CTV1 (red arrows).

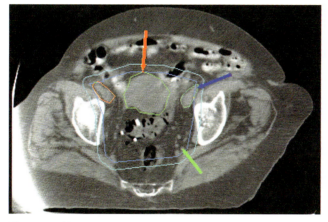

Figure 5b. Note anterior position of uterus (red arrow) and external iliac vessels (blue arrow). Posterior aspect of CTV defined by internal iliac vessels (green arrow).

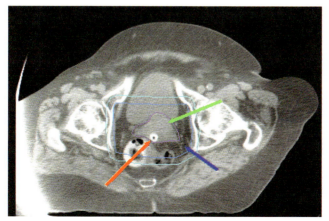

Figure 5c. Note vaginal marker (red arrow), uterus (green arrow), and parametrial tissues (blue arrow).

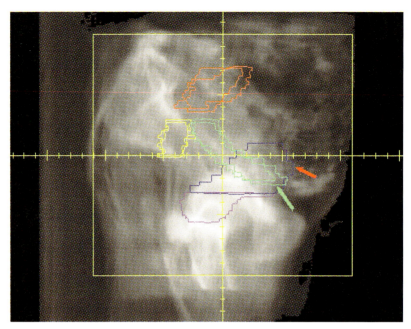

Figure 6a. Lateral DRR. Note anterior extension of uterus (red arrow) and external iliac vessels (green arrow).

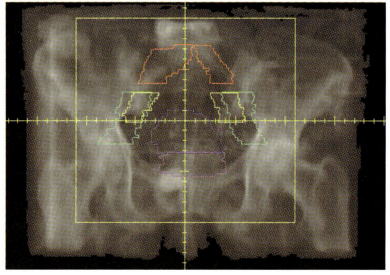

Figure 6b. Note cervix and vagina (purple), uterus (blue) common iliac vessels (orange), external iliac vessels (green), internal iliac vessels (yellow). PTV1 encompasses the CTV with a 1 cm margin.

REFERENCES

1. Boring CC, Squires TS, Tong T. Cancer statistics, 1993. CA Cancer J Clin 1993;43:7-26.

2. Morrow CP, Bundy BN, Kurman RJ, Creasman WT, Heller P, Homesley HD, Graham JE. Relationship between surgical-pathological risk factors and outcome in clinical stage I and II carcinoma of the endometrium: a Gynecologic Oncology Group study. Gynecol Oncol 1991; 40:55-65.

3. Elliott P, Green D, Coates A, Krieger M, Russell P, Coppleson M, Solomon J, Tattersall M. The efficacy of postoperative vaginal irradiation in preventing vaginal recurrence in endometrial cancer. Int J Gynecol Cancer 1994; 4:84-93.

4. Sears JD, Greven KM, Hoen HM, Randall ME. Prognostic factors and treatment outcome for patients with locally recurrent endometrial carcinoma. Cancer 1994; 74:1303-1308.

5. Perez CA, Kuske RR, Camel M, Galakatos AE, Hederman MA, Kao MS, Walz BJ. Analysis of pelvic tumor control and impact on survival in carcinoma of the uterine cervix treated with radiation therapy alone. Int J Radiat Oncol Biol Phys 1988; 14:613-621.

6. Gynecologic Oncology Group (GOG). Gynecologic Oncology Group Radiotherapy Manual, August, 1993.

7. Greer BE, Koh WJ, Figge DC, Russell AH, Cain JM, Tamimi HK. Gynecologic radiotherapy fields defined by intraoperative measurements. Gynecol Oncol 1990; 38:421-424.

8. Pendlebury SC, Cahill S, Crandon AJ, Bull CA. Role of bipedal lymphangiogram in radiation treatment planning for cervix cancer. Int J Radiat Oncol Biol Phys 1993;27:959-962.

9. Bonin S, Lanciano R, Corn B, Hogan WM, Hartz W, Hanks G. Bony landmarks are not an adequate substitute for lymphangiography in defining pelvic lymph node location for the treatment of cervical cancer with radiotherapy, 1995. In press.

10. Kim RY, McGinnis LS, Spencer SA, Meredith RF, Jennelle RLS, Salte MM. Conventional four field pelvic radiotherapy technique without computed tomography — treatment planning in cancer of the cervix: potential geographic miss and its impact on pelvic control. Int J Radiol Oncol Biol Phys 1995;31:109-112.

11. Russell AH, Walter JP, Anderson MW, Zukowski CL. Sagittal magnetic resonance imaging in the design of lateral radiation treatment portals for patients with locally advanced squamous cancer of the cervix. Int J Radiol Oncol Biol Phys 1992;23:449-455.

12. Mangan ME, Rubin SC, Rabin DS, Mikuta JJ. Lymph node nomenclature in gynecologic oncology. Gynecol Oncol 1986;23:222-226.

13. Cherry CP, Glucksmann A, Dearing K, Way S. Observations on lymph node involvement in carcinoma of the cervix. J Obstet Gynaecol Brit Emp 1953;60:368-377.

14. Twiggs LB, Potish RA, George RJ, et al. Pretreatment extraperitoneal surgical staging in primary carcinoma of the cervix uteri. Surg Gynecol Obstet 1984;158:243-250.

15. Heller PB, Malfetano JH, Bundy BN, Barnhill DR, Okagaki T. Clinical-pathologic study of stage IIB, III and IVA carcinoma of the cervix: extended diagnostic evaluation for para-aortic node metastasis. A Gynecologic Oncology Group study. Gynecol Oncol 1990;38:425-430.

16. Creasman WT, Morrow CP, Bundy BN, Homesley HD, Graham JE, Heller PB. Surgical pathologic spread patterns of endometrial cancer: a Gynecologic Oncology Group study. Cancer 1987;60:2035-2041.

17. Stehman FB, Bundy BN, Thomas G, Varia M, Okagaki T, Robert J, Bell J, Heller PB. Groin dissection versus groin radiation in carcinoma of the vulva: a Gynecologic Oncology Group study. Int J Radiation Oncol Biol Phys 1992;24:389-396.

18. Kadar N, Malfetano JH, Homesley HD. Determinants of survival of surgically staged patients with endometrial carcinoma histologically confined to the uterus: implications for therapy. Obstet Gynecol 1992;80:655-659.

19. Lanciano RM, Corn BW, Schultz DJ, Kramer CA, Rosenblum N, Hogan WM. The justification for a surgical staging system in endometrial carcinoma. Radiother Oncol 1993;28:189-196.

20. Hunt M, Schultheiss T, Desobry G, Hakki M, Hanks GE. An evaluation of setup uncertainties for patients treated to pelvic sites. Int J Radiat Oncol Biol Phys 1995. In press.

21. Emami B, Lyman J, Brown A, Coia L, Goitein M, Munzenrider JE, Shank B, Solin LJ, Wesson M. Tolerance of normal tissue to therapeutic irradiation. Int J Radiat Oncol Biol Phys 1991;21:109-122.

— *Chapter 12* —

BRAIN CANCER

Benjamin W. Corn, MD and Walter J. Curran Jr, MD

INTRODUCTION

Brain tumors arise in 15,000 patients annually in the United States.[1] Malignant gliomas (glioblastoma multiforme, GBM) and astrocytoma with anaplastic foci (AAF) comprise 30% of the primary intracranial brain tumors that are diagnosed in adults. Since malignant glioma is the dominant primary brain tumor encountered in adults, the discussion will focus on the role of conformal techniques for the treatment of this challenging category of brain tumors.

Malignant gliomas typically present as large, rapidly growing lesions arising in the cerebral hemispheres. Although surgical resection represents the initial treatment for many gliomas,[2] radiation therapy continues to have an important role in the management of both completely and incompletely resected lesions. Moreover, radiation therapy represents the most important cytotoxic agent in the treatment of malignant gliomas since these tumors are relatively insensitive to chemotherapy and do not tend to metastasize. Therefore, local control of tumor growth could theoretically translate into cure.

This chapter will focus on the treatment of gliomas with conformal external beam radiotherapy. Although interstitial brain implants and stereotaxic irradiation techniques have emerged as potentially valuable options, these strategies are beyond the scope of the present discussion.

ANATOMY

Traditionally quoted[3] dimensions of the adult human brain are 16 cm anteroposteriorly, 14 cm transversely and 12 cm rostrocaudally. The volume of the human brain approximates 1,300 cm^3 while the surface area is approximately 2,000 cm^2. The average thickness of the cerebral cortex is 2.5 mm. As for all sites discussed in this textbook, a range of dimensions exists in the population, underscoring the importance of individualizing treatment planning for each patient.

The brain is the most important human organ and is protected by the surrounding cranium. The skull contains several radiographic and surface topographic points of reference (Figure 1). The sella turcica is centrally located and delimits the pituitary and the adjacent optic chiasm. The most caudal part of the anterior cranial fossa is the cribriform plate. Other

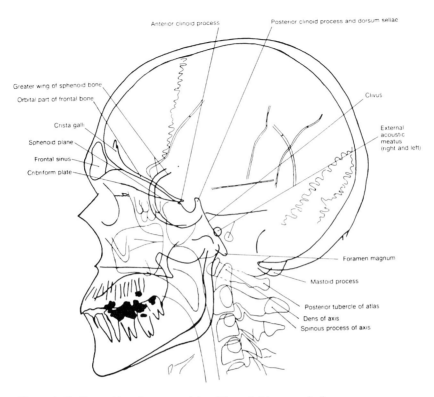

Figure 1. Radiographic reference points of the adult human skull.

easily identifiable structures on lateral radiographs include the clivus and the external auditory meatus. Although various radiographic pearls[4] relating other neuroanatomical structures (e.g., tentorium cerebelli, hypothalamus, pineal body) to these landmarks have been published, the presently available imaging technologies (e.g., CT and MRI) and their interfacing three-dimensional software packages have rendered many of these guidelines obsolete. As such, the radiation treatment planning of all brain tumors should be planned de novo and the time-honored landmarks should be used primarily to help confirm beam localization.

PATTERNS OF SPREAD

The volume of brain that is included in the radiation treatment portal depends on the geometry of the specific tumor, the clinician's knowledge of its biology and the potential for intracranial dissemination. In a comprehensive review, Sheline[5] summarized the controversy surrounding the use of whole brain irradiation (WBI) versus partial brain irradiation (PBI) for malignant glioma. In separate autopsy-based reports, Concannon[6] and Kramer[7] referred to the diffuse failure patterns associated with PBI of malignant glioma and therefore advocated that the entire brain be considered part of the target. Salazar et al.[8] described an experience of 148 malignant glioma patients irradiated with varying techniques at the University of Rochester. A survival advantage was noted for those whose irradiation portals exceeded 200 cm^2 in comparison to those who received treatment through smaller beam apertures. Similarly, Todd et al.[9] note a survival advantage among those who were treated through portals >500 cc in comparison to those irradiated to smaller volumes. These studies were reported in an era when neuroimaging capabilities were unable to reliably define tumor volumes and corresponding radiation fields. Accordingly, it is possible that large portal treatment (e.g., WBI) was merely a surrogate for the inclusion of the entire lesion within the treated volume. Moreover, those investigators did not rigorously control for other treatment factors (e.g., dose[10]) that may impact on control of glioblastoma multiforme. In randomized trials, Payne et al.[11] and Ramsey et al.[12] observed no outcome differences when comparing glioma patients treated by partial fields to those treated by whole brain fields. Although these trials may be criticized because of their small size, several large retrospective experiences that were predicated on the use of PBI contained comparable survival to that obtained from WBI. The weight of evidence therefore suggests that outcome is not compromised from the use of smaller fields. When this phenomenon is coupled with the potential preservation of function and cognition associated with smaller fields, the argument for subtotal brain irradiation is strengthened.

Several lines of evidence have been published to determine the microscopic limits of tumor extension. In a classic paper, Hochberg and Pruitt[13] evaluated CT scans performed no more than two months prior to post-mortem examination and determined that in 29 of 35 glioblastoma multiforme patients evaluated, gross and microscopic tumor extent fell within a 2 cm rim of the primary lesion as identified on the initial diagnostic scan. Since this was only a two-dimensional analysis, it is not clear that additional patients with disease extending beyond the 2 cm limits on axial sections would not have been found to have additional disease present on sagittal, coronal, or other anatomical sections.

At the Mayo Clinic, Kelly et al.[14] obtained consent from glioma patients to perform serial stereotaxic biopsies in a trajectory that extended outside abnormalities identified on CT or MRI scans. Positive biopsies were obtained from sixteen out of twenty-three patients in ostensibly normal T1-weighted regions and four out of ten patients in purportedly normal T2-weighted regions. That study, however, did not relate the probability of obtaining a positive biopsy to the distance between the imaging abnormality and the site of the positive biopsy.

Burger et al.[15] correlated whole brain histologic sections of 11 untreated GBM patients with the final CT scan obtained prior to death. Tracings of the pathologic extent of tumor were compared to tracings of the radiologist's impression of tumor on the corresponding brain slices. In all 11 cases, the contrast enhancing lesion fell short of delimiting the entire pathologic extent of the neoplasm. An additional 3 cm rim would have encompassed all tumor cells in 9 of 11 cases. Similar observations were reported by Wallner et al.[16] and Halperin et al.[17]

In a recent summary of the experience from Johns Hopkins University, Wie and Grossman[18] sounded a cautionary note by observing that tumor failure along the white matter tracts could be a problem especially among patients with frontal or occipital lesions. If this experience is confirmed by others, it may be necessary to reassess the classical teaching regarding the radiation planning, at least for lesions arising in those locations. In the meantime, however, the overwhelming priority for malignant glioma is to address local tumor clonogens before targeting the more peripheral cells.

In synthesizing the controversy surrounding WBI versus PBI for malignant glioma, Halperin[19] articulated the two antipodean conclusions that could reasonably be drawn from the literature. First, that WBI must be used (for at least part of the treatment) since WBI represents the only means of confidently covering the entire tumor volume. Alternatively, that PBI is preferable since it is associated with equivalent albeit poor outcome. The philosophy at Fox Chase Cancer Center has been to adopt the latter approach since WBI appears to be more morbid. Moreover, since local recurrence remains a consistently reported pattern of failure in all series, it is necessary to concentrate on this as yet unremedied problem prior to attending to the theoretical issue of distal failure. Finally, since a subset of malignant glioma patients may actually have local rather than diffuse disease[20] and all patients have the highest density of tumor cells in the enhancing volume, conformal irradiation strategies offer an opportunity to escalate dose in the hopes of safely improving tumor control in more patients.

DIAGNOSTIC STUDIES

The imaging of brain neoplasms has undergone a dramatic revolution during the past two decades. Although CT ushered in an era of unique axial anatomical representation of neuroanatomy and neuropathology, MR imaging has supplanted CT as the imaging modality of choice for the central nervous system.[21] Lesions are thus best visualized on MR scans, yet even with the present generation of scanners, the microscopic extent of disease is not directly identifiable. Moreover, most 3D radiation treatment planning systems are predicated on CT software. Therefore, presently available planning capabilities are unable to harness the best available diagnostic technology.

At present, many departments of radiation oncology that offer 3D capability follow the relatively crude procedure of attempting to obtain treatment planning scan slices at precisely the same interval and orientation as the baseline diagnostic scan. Alternatively, geometric correlations can be made between images from baseline and treatment planning scans which were not taken in the same plane. This process is both time consuming and inaccurate. Ideally, it should be possible to combine the complementary information from separate diagnostic modalities (e.g., CT and MRI) into a single coherent study. Such an approach is being explored by investigators at the Harvard Joint Center.[22] An interface is also being developed by the Picker Corporation for the ACQSIM CT Simulator to allow the importation of MR data. Until this image fusion technology is upgraded, however, a significant limitation to transpose baseline diagnostic information into useful treatment planning data remains. In the meantime, it will be essential that the entire neuro-oncologic team (including neurosurgeons, neuroradiologists and radiation oncologists) provide input into case management at the earliest possible juncture so that effective treatment plans can be formulated.

IMMOBILIZATION

Proper immobilization techniques are necessary to permit the use of small margins around the clinical target volume so that normal tissue dose and associated complications can be reduced. Aside from toxicity considerations, inattention to immobilization can compromise tumor control probability. It is difficult to properly align the head in anticipation of complex radiation treatment. Therefore, supplemental immobilization of the head is essential to assure reproducibility of set up and to allow irradiation of the target with blocking of normal structures. At Fox Chase Cancer Center we have employed two types of fixation devices, customized mouth pieces that are attached to the table frame and Aquaplast net masks (W.F.R. Corporation, Smithers, New Jersey). We currently favor laser alignment in conjunction with these contoured masks as our optimum positioning technique. In a report from the Massachusetts General Hospital, Verhey et al.[23] showed that such an approach provided immobilization at the level of 2.2 mm ± 1.4 mm SD. Those authors therefore postulated that, assuming a normal distribution of errors, one could expect to be within 5 mm of the desired position in 95% of patient treatments, probably adequate for most high energy photon plans for the central nervous system.

PATIENT POSITIONING

Although creative patient positioning can obviate excessive irradiation of critical structures, the capacity of a patient to cooperate with changes in head positioning may limit the ability to achieve the desired set up. For instance, a simple approach towards irradiation of frontal lesions without traversing the optic apparatus entails head flexion in anticipation of an anterior and lateral wedge beam pair (see Figure 2). At times, however, the cancer patient does not have sufficient flexibility in the neck to properly tuck the head away from the approaching beam. To circumvent this problem, one may consider rotating the table by 90° and angling the beam in the sagittal plane of the body. Ideally, the clinician should give consideration to patient positioning prior to initiation of the CT simulation.

LOCALIZATION

Definition of Target Volumes

As alluded to above, there has been a general evolution in the high-grade glioma literature from WBI (1960s), to WBI plus partial brain boost (1970s), to initial fields encompassing the tumor plus edema prior to cone-downs to the tumor itself (1980s), and most recently to initial fields encompassing the T2 magnetic resonance image and cone-downs to the gadolinium-enhancing lesions on T1-weighted scans. This progression has been reflected in the treatment guidelines of glioma protocols during the past three decades.

Most malignant gliomas are subtotally resected[24] and the gross tumor volume (GTV) is determined from the preoperative scan. A unique concern that arises in the treatment planning of malignant gliomas is the uncertain volume redistribution of brain parenchyma and cerebrospinal fluid within the cranial vault after resection has been completed. Although a similar dilemma is theoretically present in the planning of boost volumes in the treatment of breast cancer patients following lumpectomy, that problem is alleviated by the placement of surgical clips to delimit the tumor bed. Clips can only be used with caution for the brain tumor population where interval MR scanning is the norm. Even if nonmagnetic clips can be inserted by the neurosurgeon, there is concern that such clips may be subject to migration.

At Fox Chase Cancer Center, our approach to malignant glioma patients who are not entered onto protocols has been to treat an initial field to 46 Gy, followed by a cone-down boost to deliver an additional 14 Gy (fraction size, 2 Gy). The gross tumor volume is derived from the gadolinium-enhancing regions on T1-weighted MRI scans. The clinical tumor volume (CTV) must encompass the gross tumor volume plus the presumed subclinical extension of disease. The initial CTV is taken as the region of abnormality on T2-weighted MR scans while the cone-down CTV is taken as the gadolinium-enhancing region on T1-weighted MR scans. Determination of the CTV is also modified by the proximity of critical structures (e.g., optic apparatus) and anatomic partitions such as the base of skull that are thought to represent barriers to tumor invasion. The planning target volume (PTV) in each case is taken to be the

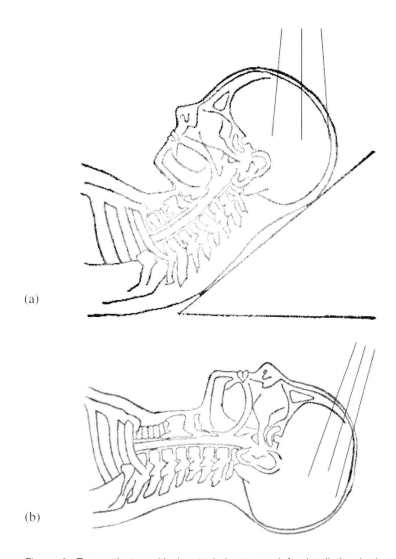

Figure 2. Two patient positioning techniques used for irradiating brain tumors. (a) The head is tucked so as to allow treatment with an anterior field while shielding the optic apparatus. (b) The head remains in a neutral position, while the couch and gantry are rotated to avoid the optic apparatus.

corresponding CTV plus 5 mm margin to account for set up uncertainty. We have postulated that set up uncertainty for brain tumor patients is similar to the published experience[25,26] for patients with head and neck cancer.

Definition of Critical Structures

A fundamental concern of the radiation oncologist in designing treatments for glial tumors is the sensitivity of the surrounding normal brain tissue to the beam. Potential late complications include cerebral necrosis, neurocognitive deficits, optic injury, and dysfunction of the pituitary-hypothalamic axis. Several estimates of central nervous system tolerance to irradiation have been proposed. Rubin speculated that a dose of 50 Gy to the whole brain is associ-

Table 1. Target Volumes: High Grade Glioma

	Initial Field	Cone-Down Field
GTV	Gross tumor volume (gadolinium enhancement on T1 weighted MR image)	Gross tumor volume (gadolinium enhancement on T1 weighted MR image)
CTV	Gross tumor volume + edema (T2 MRI abnormality)	Gross tumor volume
PTV	$CTV_{initial}$ + set up error (5 mm)	CTV_{boost} + set up error (5 mm)

ated with a 5% risk of necrosis.[27] Marks et al.[28] reported a 5% incidence of radiation necrosis among patients who received total doses above 55 Gy at fractions ≥2.0 Gy in contrast to no risks encountered among those treated to 54 Gy in daily fractions of ≤1.8 Gy. Sheline[5] argued that even 60 Gy of PBI could be pursued with reasonable safety at conventional fractionation. After reviewing the literature, Parsons suggested that a dose of 54 Gy can be delivered to the optic apparatus with only a low likelihood of ensuing blindness and optic neuropathy.[29] Constine et al.[30] observed a 65% level of gonadal dysfunction and a 74% incidence of hypothyroidism after a mean pituitary dose of 53.60 Gy was administered to a group of 32 pediatric and adult patients who consented to neuroendocrine assessment 2 to 13 years after receiving cranial radiotherapy for non-pituitary brain tumors. Since uncertainty still surrounds the precise radiation tolerance of the central nervous system it is axiomatic that efforts to minimize dose deposition within the normal brain are essential.

CT SCAN LOCALIZATION OPTIMIZATION & VIRTUAL SIMULATION

The actual scanning protocol should be optimized based on considerations of volume averaging, target and normal structure geometry, procedure time and image quality, particularly of digitally reconstructed radiographs (DRRs). When planning brain irradiation at Fox Chase Cancer Center (Table 3), it has been our policy to obtain CT images that extend from the top of the cranium through the inferior level of the second cervical vertebral body. Slice thickness is 2 mm, and slice separation is 2 mm. At the discretion of the physician, patients receive non-iodinated contrast material 30 minutes prior to scanning to decrease the risk of

Table 2. Radiation Tolerance (TD 5/5) of Representative CNS Structures*

Organ	Endpoint	Dose (Gy)
Partial brain	Necrosis, infarction	60
Optic nerve, chiasm	Blindness	50
Retina	Blindness	45
Brainstem	Necrosis, infarction	60 (partial brainstem) 50 (whole brainstem)

*Modified after Emami et al.[31]

untoward reactions. We have found this scanning protocol to yield high quality CT images and DRRs.

Treatment localization and CT simulation procedures for the brain are performed by the therapist using the procedures outlined in Chapter 3. After the CT scan of the patient is complete and transferred to the virtual simulator, the isocenter is calculated in the treatment localization mode using the CTV as defined by the physician. The isocenter as defined by the intersection of the lateral and sagittal wall lasers are then tattooed onto the patient, after which the patient then leaves the CT simulation suite. The radiation oncologist outlines critical structures such as optic nerves and chiasm. Beam arrangement optimization is then carried out by the dosimetrist and/or physician. DRRs of the planned beam fields with the target and other structures are obtained and conformal beam blocks are drawn using the treatment planning guidelines following.

An especially attractive feature of CT simulation of brain irradiation is that DRRs can be created for fields that cannot be simulated despite the fact that such fields can be treated on the linear accelerator where portal verification *can* be carried out. For instance, many of the oblique fields that are desirable for the treatment of brain tumors cannot be captured on a simulation radiograph. This is problematic because it does not allow the oncologist to spatially visualize the tissues traversed by the radiation beam and consequently compromises the design of protective customized blocks. In contrast, DRRs allow visualization of the beam appearance and permit the clinician to determine if the beam is encompassing the tumor or moving too close to a critical structure that needs to be shielded.

TREATMENT PLANNING

Several time-honored beam arrangements have been used in the radiation treatment of brain tumors. Lateral opposed beams are commonly employed to treat tumors arising in the posterior-frontal or anterior-parietal lobes in deference to ease of set up and verification; however, such portal arrangement is not ideal because a large volume of cerebral cortex is invariably exposed to moderately high doses of irradiation. Frontal lesions are often approached with a "wedge-pair" of beams (e.g., orthogonally opposed lateral and anterior beams); however, care must be taken to minimize irradiation of the optic apparatus by properly positioning the patient on the treatment couch (see "Immobilization," pp. 152-153). Similar concerns arise when treating posterior-parietal or occipital tumors with posterior and lateral isocentric perpendicular beams. Lesions occupying the lateral aspect of one of the temporal lobes can often be treated with stationary vertex and lateral portals or arc rotations; however, when the bulk of disease is located in the medial aspect of the temporal lobes, the conventional wisdom defaults to lateral opposed beams that increase the risk of temporal lobe necrosis.[32] The use of

Table 3. Features of CT Treatment Planning for Malignant Gliomas

- Scanned region: from top of cranium (cephalad) to bottom of C2 (caudad)
- CT slice thickness: 2 mm
- CT slice separation: 2 mm
- IV contrast: non-iodinated agents administered 30 minutes prior to scanning
- Immobilization: contoured net masks (Aquaplast)
- Head position: individualized to location of tumor and critical structures

CT/virtual simulation opens the possibility of non-coplanar beam arrangements similar to those used in stereotactic irradiation. By developing conformal plans which use non-coplanar beams, beam overlap regions can be minimized, thereby decreasing the dose to normal brain tissue. As this technology becomes more widely disseminated, a major step forward will have been taken towards the optimization of the therapeutic index.

TREATMENT VERIFICATION

DRRs for patients with malignant glioma can be used to verify field arrangements by using bony landmarks. Conventional simulation is not needed and as mentioned some beam arrangements which can not be depicted by conventional simulation (e.g., vertex fields) can be readily demonstrated by CT simulation.

CASE ILLUSTRATIONS

Two cases are presented to illustrate some of the aforementioned points. Figure 3 shows a 52-year-old man who presented with frontal lobe headaches. Subtotal debulking was carried out and the histopathology was signed out GBM. Following surgery the patient still had some mass effect present on CT scan and he developed new facial droop. As such, the decision was made to initiate PBI promptly with opposed lateral portals (Figure 3a). After the delivery of 20 Gy through these beams, the field arrangement was switched to lateral portals with a 40° superior oblique field for the next 30 Gy (Figure 3b). As the optic chiasm was still irradiated to full dose, the final cone-down was with lateral portals and a 20° oblique field that was completely off the chiasm (Figure 3c). DRRs were particularly useful for imaging the CTV (stippled volume) and the surrounding normal structures on the unconventional views. Often beams such as the one depicted in Figure 3c cannot be obtained at the time of simulation because the image intensifier would hit the table. By generating DRRs, one can be certain of the location of the optic chiasm and fashion blocks accordingly.

Figure 4 depicts a 65-year-old man with recurrent oligodendroglioma. The initial plan was to treat the patient in a chin-tucked position with anterior and lateral beams. At the time of simulation, however, the patient could not sufficiently flex his neck. The CT simulator allowed the physicians to achieve the same goal by using lateral beams in conjunction with a superior oblique portal. As such, this second case attests to the flexibility that CT simulator technology adds to the system of treatment planning.

CONCLUSION

Conformal irradiation of glial tumors is predicated upon many of the principles of classic radiotherapy. Immobilization requirements for conformal irradiation are often more stringent than those associated with conventional x-ray treatment since many of these patients are selected for conformal strategies in order to more safely escalate dose, or because of the proximity of the target to important critical structures in the central nervous system. Conformal plans and CT-based simulation facilitate the design of portals that restrict the volumes of normal brain irradiated without sacrifice of clinical benefits. It is hoped that the creative imple-

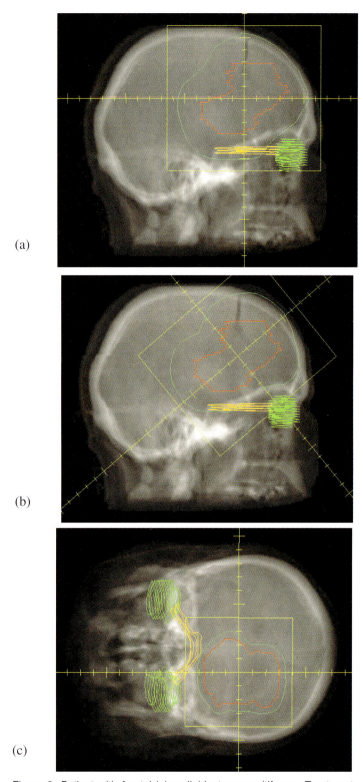

Figure 3. Patient with frontal lobe glioblastoma multiforme. Treatment was expedited with lateral portals (a) prior to the planning of more sophisticated beam arrangement (b,c) as described in text.

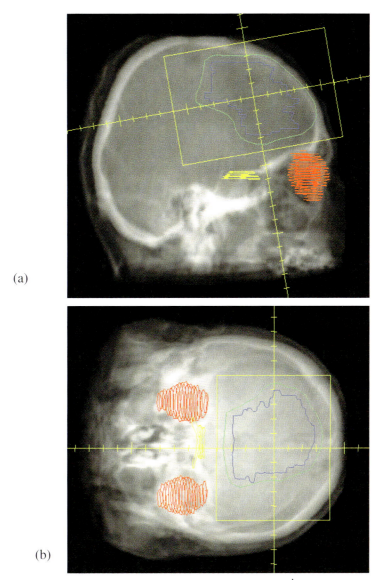

Figure 4. Lateral (4a) and superior oblique (4b) portals of patient with recurrent oligodendroglioma.

mentation of conformal approaches will increase the willingness of physicians to relinquish the nihilistic philosophies that stand in the way of overcoming a disease that will require aggressive solutions if it is to be conquered.

REFERENCES

1. Levin VA, Gutin PH, Leibel SA. Cancer: Principles and Practice, 4th edition. Philadelphia, Pennsylvania: J.B. Lippincott, 1993.

2. Simpson JR, Horton J, Scott C, et al. Influence of location and extent of surgical resection on survival of patients with glioblastoma multiforme: results of three consecutive radiation ther-

apy oncology group (RTOG) clinical trials. Int J Radiat Oncol Biol Phys 1993;26:239-244.

3. Karlsson UL, Leibel SA, Wallner K, Davis LW, Brady LW. Principles and Practice of Radiation Oncology, 2nd edition. Philadelphia: JB Lippincott, 1992.

4. Wicke L. Atlas of Radiologic Anatomy. Munich: Urban and Schwarzenberg, 1987.

5. Sheline GE. Radiation therapy for primary tumors. Semin Oncol 1975;2:29-42.

6. Concannon JP, Kramer S, Berry R. The extent of intracranial gliomata at autopsy and its relationship to techniques used in radiation therapy of brain tumors. Am J Roentgenol 1960;84:99-107.

7. Kramer S. Tumor extent as a determining factor in radiotherapy of glioblastomas. Acta Radiol 1969;8:111-117.

8. Salazar OM, Rubin P, McDonald JV, Feldstein ML. Patterns of failure intra-cranial astrocytomas after irradiation: analysis of dose and field factors. Am J Roentgenol 1976;126:279-292.

9. Todd ID. Choice of volume in x-ray treatment of supratentorial gliomas. Br J Radiol 1963;36:645-649.

10. Green SB, Shapiro WR, Burger PC, et al. A randomized trial of interstitial radiotherapy boost for newly diagnosed malignant glioma: BCTG trial 8701. Proc Amer Soc Clin Oncol 1994;174.

11. Payne DG, Simpson WJ, Keen C, et al. Malignant astrocytoma: hyperfractionated and standard radiotherapy with chemotherapy in a randomized prospective clinical trial. Cancer 1982;50:2301-2306.

12. Ramsey RS, Brand WN. Radiotherapy of glioblastoma multiforme. J Neurosurg 1973;39:197-202.

13. Hochberg FH, Pruitt A. Assumptions in the radiotherapy of glioblastoma. Neurology 1980;30:907-911.

14. Kelly PJ, Daumas-Duport C, Scheithauer BW, et al. Stereotactic histologic correlations of computed tomography and magnetic resonance imaging defined abnormalities in patients with glial neoplasms. Mayo Clin Proc 1987;62:450-459.

15. Burger PC, Heinz ER, Shibata T, et al. Topographic anatomy and CT correlations of in the untreated glioblastoma multiforme. J Neurosurg 1988; 68:698-704.

16. Wallner KE, Galicich JH, Krol G, et al. Patterns of failure following treatment for glioblastoma multiforme and anaplastic astrocytoma. Int J Radiat Oncol Biol Phys 1989;16:1405-1409.

17. Halperin EC, Bentel G, Heinz ER, Burger PC. Radiation therapy planning in supratentorial glioblastoma multiforme: an analysis based on post mortem topographic anatomy with CT correlations. Int J Radiat Oncol Biol Phys 1989;17:1347-1350.

18. Wie BA, Grossman SA, Wharam MD, Chun M. Dissemination of high grade astrocytomas (HGA) along white matter tracts (WMT): observations and therapeutic implications. Proceedings of ASCO1993;12:177.

19. Halperin EC, Burger PC, Bullard DE. The fallacy of the localized supratentorial malignant glioma. Int J Radiat Oncol Biol Phys 1988;15:505-509.

20. Loeffler JS, Shrieve DC, Alexander E. Radiosurgery for glioblasoma multiforme: the importance of selection criteria. Int J Radiat Oncol Biol Phys 1994;30:731-733.

21. Bragg DG, Osborn AG. CNS imaging of neoplasms. Int J Radiat Oncol Biol Phys 1991;21:841-845.

22. Kooy HM, Van Herk M, Barnes P, et al. Image fusion for stereotactic radiotherapy and radiosurgery treatment planning. Int J Radiat Oncol Biol Phys 1994;28:1229-1234.

23. Verhey L, Goitein M, McNulty P, et al. Precise positioning of patients for radiation therapy. Int J Radiat Oncol Biol Phys 1982;8:289-294.

24. Curran WJ, Scott CB, Horton J, et al. The influence of tumor size and location and extent of surgery on outcome for anaplastic astrocytoma: a report from three Radiation Therapy Oncology Group (RTOG) trials. J Neuro-Oncol 1992;12:219-227.

25. Hunt MA, Kutcher GJ, Burman C, Fass D, Harrison L, Leibel S, Fuks Z. The effect of setup uncertainties on the treatment of nasopharynx cancer. Int J Radiat Oncol Biol Phys 1993;27:437-447.

26. Huizenga H, Levendaag PC, De Porre PM, Visser AG. Accuracy in radiation field alignment in head and neck cancer: a prospective study. Radiotherapy and Oncology 1988; 11:181-188.

27. Rubin P, Casarett G. Clinical radiation pathology. Philadelphia, Pennsylvania: WB Saunders, 1968.

28. Marks JE, Baglan RJ, Prassad SC, et al. Cerebral radionecrosis: incidence and risk in relation to dose, time, fractionation and volume. Int J Radiat Oncol Biol Phys 1982;8:37-43.

29. Parsons JT, Fitzgerald CR, Hood CI, et al. The effects of irradiation on the eye and optic nerve. Int J Radiat Oncol Biol Phys 1983;9.

30. Constine LS, Woold PD, Cann D, et al. Hypothalamic-pituitary dysfunction after radiation for brain tumors. N Engl J Med 1993;328:87-94.

31. Emami B, Lyman J, Brown A, Coia L, Goitein M, Munzenrider JE, Shank B, Solin L, Wesson M. Tolerance of normal tissue to therapeutic irradiation. Int J Radiat Oncol Biol Phys 1991;21:109-122.

32. Cooley G, Gillin MT, Murray KJ, Wilson JF, Janjan NA. Improved dose localization with dual energy photon irradiation in treatment of lateralized intracranial malignancies. Int J Radiat Oncol Biol Phys 1991;20:815-821.

Chapter 13

PROSTATE CANCER

W. Robert Lee, MD and Gerald E. Hanks, MD

INTRODUCTION

Adenocarcinoma of the prostate is the most common noncutaneous malignancy diagnosed in American men. Over 200,000 new cases will be diagnosed in 1994 and the incidence has been increasing over the past decade. External beam radiation therapy occupies a central position in the treatment of this malignancy. At Fox Chase Cancer Center all patients with prostate cancer who elect to be treated with definitive radiation therapy are planned and treated using three dimensional CT-based conformal techniques. This conformal method has resulted in improved rates of cancer control when compared to conventional treatment methods.[1]

ANATOMY

The prostate is generally described as a walnut-shaped organ that typically weighs 20 grams at maturity. The prostate is found posterior to the symphysis pubis and surrounds the prostatic urethra. The gland is bounded by a fibromuscular band of transversely oriented fibers that blends with the prostatic stroma beneath it and as such a true capsule in the strict sense does not exist. It is separated from the rectum by the rectovesical septum. The seminal vesicles lie superior to the gland and join with the ampullary end of the ductus deferens to form the ejaculatory duct which empties into the prostatic urethra. McNeal has described five histologically distinct zones: the anterior zone; the peripheral zone (approximately 75% of the glandular component); the central zone; the preprostatic urethra and the transitional zone.[2] Carcinoma of the prostate arises in the peripheral zone in the majority of cases and it is this portion of the gland which is palpable on digital rectal exam. The 25% of cancers that arise in the transition zone appear to have a different natural history and are usually diagnosed incidentally at the time of TUR.

PATTERNS OF SPREAD

Step-sectioned whole organ specimens from radical prostatectomies have provided an accurate picture of the local spread patterns of adenocarcinoma of the prostate. The local patterns of spread are largely determined by the site of origin of the particular cancer.[3] Cancers arising in the transitional zone seldom penetrate the capsule except anterolaterally and at the apex. Peripheral zone cancers penetrate the capsule much more frequently, preferentially along the posterolateral angle of the prostate near the neurovascular bundle. The incidence of capsular penetration is strongly correlated to tumor volume.

Radical prostatectomy specimens have also provided valuable information about the risk of seminal vesicle invasion by cancer. Tumor involvement of the seminal vesicles has been associated with high tumor grade, advanced clinical stage, large tumor volume and location of the primary tumor.

The lymphatic drainage of the prostate is primarily to the obturator and internal iliac (hypogastric) lymph node groups. Less commonly the external iliac, common iliac, presacral, presciatic and paraaortic nodes may be involved. The risk of lymph node metastases is closely correlated with tumor volume, pretreatment PSA level (a surrogate for tumor burden) and histologic grade.[4]

DIAGNOSTIC STUDIES/STAGING

All patients with prostate cancer should have a serum PSA test, radionuclide bone scan and digital rectal examination prior to initiating definitive radiotherapy. At Fox Chase we have routinely ordered MRI scans of the prostate using an endorectal or Helmholtz coil to determine whether there is radiographic evidence of spread beyond the gland (capsular penetration, seminal vesicle involvement). In patients to be treated with external beam radiotherapy CT scans of the abdomen are not sufficiently sensitive or specific to accurately stage patients with prostate cancer and therefore we have not recommended this test unless performed as part of the treatment planning process in which the study has some value. At Fox Chase patients are staged using the most recent AJCC TNM system shown in Table 1.

IMMOBILIZATION AND PATIENT POSITIONING

All patients with prostate cancer are treated in the supine position with arms crossed on the chest. A custom-made Alpha Cradle cast extending from the mid-thoracic spine to the mid-thigh in addition to a foot holder are used throughout the course of treatment to improve set up reproducibility and reduce external rotation of the hips. Patients are encouraged to receive each treatment with the bladder as full as possible.

CT SCAN LOCALIZATION OPTIMIZATION

All patients are scanned from the L4-5 interspace to 5 cm below the ischial tuberosities. The current protocol utilizes 3 mm slice thicknesses at 5 mm intervals. In all patients, a retrograde urethrogram (Reno-M-30, 10 mL) and intravenous contrast (Isovue 200, 25 mL) are administered at the time of CT simulation. The DRRs obtained with these scanning parameters are of high quality and there is no need for conventional simulation to obtain a film to allow for portal verification.

Table 1. AJCC TNM Classification System for Prostate Cancer

Primary Tumor (T)
- TX: Primary tumor cannot be assessed
- T0: No evidence of primary tumor
- T1: Clinically inapparent tumor
  - T1a: Incidental histologic finding in ≤5% of resected tissue
  - T1b: Incidental histologic finding in >5% of resected tissue
  - T1c: Tumor diagnosed by elevated PSA prompted biopsy
- T2: Tumor confined to the prostate
  - T2a: Tumor involves ≤1/2 lobe
  - T2b: Tumor involves >1/2 lobe, but not both lobes
  - T2c: Tumor involves both lobes
- T3: Tumor extends through prostatic capsule
  - T3a: Unilateral extracapsular extension
  - T3b: Bilateral extracapsular extension
  - T3c: Tumor invades seminal vesicle(s)
- T4: Tumor is fixed or invades adjacent structures other than seminal vesicles
  - T4a: Tumor invades any of: bladder neck, external sphincter, rectum
  - T4b: Tumor invades levator muscles and/or is fixed to pelvic wall

Regional Lymph Nodes (N)
- NX: Regional lymph nodes cannot be assessed
- N0: No regional lymph nodes metastases
- N1: Metastasis in a single lymph node ≤2 cm in greatest dimension
- N2: Metastasis in a single lymph node >2 cm but ≤5 cm in greatest dimension or multiple lymph node metastases, none greater than 5 cm in greatest dimension
- N3: Metastasis in a lymph node >5 cm in greatest dimension

Distant Metastasis (M)
- MX: Presence of metastasis cannot be assessed
- M0: No distant metastasis
- M1: Distant metastasis present
  - M1a: Nonregional lymph node(s)
  - M1b: Bone(s)
  - M1c: Other site(s)

TARGET LOCALIZATION

Definition of Gross Tumor Volume

In the majority of cases the prostate is easily identified on axial CT images. The most difficult area to visualize is the apex of the gland. At Fox Chase all patients undergo retrograde urethrography and the apex of the prostate gland is estimated to be 1.0 centimeter above the cone of the contrast column based on comparison to MRI images. Intravenous contrast (Isovue 200, 25 mL) is also utilized immediately prior to the CT simulation procedure which helps to delineate the bladder from the prostate gland. The gross tumor volume for prostate cancer treatment includes the prostate gland and occasionally the seminal vesicles. The seminal vesicles can be identified posterolaterally (Figure 1).

Definition of Clinical Target Volume

The clinical target volume represents the gross tumor volume (GTV) plus all tissues which are thought to contain microscopic disease. In patients with low stage, low PSA, and low Gleason grade the risk of extraprostatic disease is low and the GTV and CTV are equivalent. In patients with locally advanced prostate cancer or those with a pretreatment PSA of greater

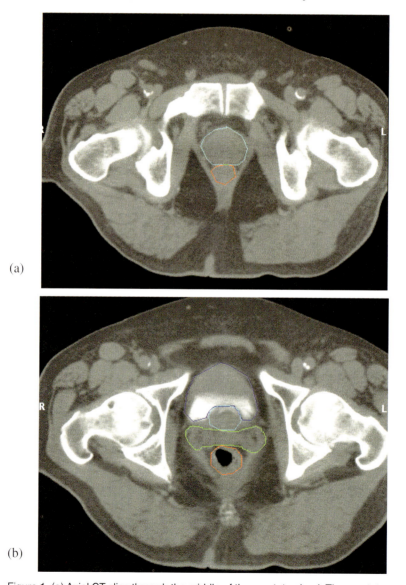

Figure 1. (a) Axial CT slice through the middle of the prostate gland. The prostate gland and rectum are outlined. (b) Axial CT slice through the superior portion of the prostate gland. The prostate gland, seminal vesicles, bladder and rectum are outlined.

than 10 ng/mL, the CTV may include the seminal vesicles and the pelvic lymph nodes. At Fox Chase the pelvic lymph nodes are electively irradiated if the risk of metastases to these nodes is felt to be greater than 15%. The treatment guidelines according to clinical stage, Gleason grade, and pretreatment PSA used at Fox Chase are outlined in Table 2.

Definition of Planning Target Volume

The planning target volume (PTV) encompasses the CTV plus a margin to account for set up uncertainty and organ motion. Studies at Fox Chase on immobilized prostate cancer patients have determined that a margin of 1 cm (1.5 cm when the lymph nodes are included in the CTV) in all directions is adequate to cover the CTV with a margin of 3 millimeters in over 95% of cases if prostate motion is not taken into account. The magnitude of prostate motion during treatment is not well studied at present and not specifically accounted for in our technique. Therefore, 1 cm is added to the CTV in all directions to define the PTV. To allow for adequate buildup the margins of the cerrobend blocks are an additional 5 mm from the PTV, so that the aperture margin is 1.5 centimeters beyond the CTV. When these guidelines are followed the PTV is enclosed by the 95% isodose line in virtually all cases.

DEFINITION OF CRITICAL STRUCTURES

Critical structures in the treatment of prostate cancer include the rectum, bladder and, in patients receiving elective pelvic lymphatic irradiation, the small bowel. The published toler-

Table 2. Fox Chase Treatment Guidelines for Clinically Localized Prostate Cancer

T Stage	PSA (ng/mL)	Treatment volumes*	Dose (cGy)
T1, T2a/b	<10	GTV = PR CTV = PR PTV = CTV + 1 cm	PTV = 68-70 Gy
T1, T2a/b	>10 or Gleason ≥7	GTV = PR CTV1 = PR + SV + LN CTV2 = PR PTV = CTV + 1 cm (1.5 cm around LN)	PTV1 = 45-50 Gy PTV2 = 68-72 Gy
T2c, T3	Any	GTV = PR ± SV CTV1 = PR + SV + LN CTV2 = PR + SV CTV3 = PR PTV = CTV + 1 cm (1.5 cm around LN)	PTV1 = 45-50 Gy PTV2 = 57-60 Gy PTV3 = 68-72 Gy

*GTV indicates gross tumor volume; CTV, clinical target volume; PTV, planning target volume; PR, prostate; SV, seminal vesicles; LN, pelvic lymph nodes.

ance doses of these organs are listed in Table 3 but it must be remembered that these tolerance doses represent estimates based largely on anecdotal reviews and the tolerance doses for partial organ irradiation are even more speculative.[5] Early data from the conformal experience at Fox Chase suggests that incidence of Grade III to IV rectal complications in the 70 to 75 Gy range may be higher than expected using the published guidelines.[6]

TREATMENT PLANNING

The guidelines for radiation dose prescriptions are shown in Table 2. Patients in whom the pelvic nodes are irradiated usually receive a dose of 45 to 50 Gy to these nodes. One or two reductions in the treated volume are made to include the seminal vesicles and prostate or the prostate alone to doses of 68 to 72 Gy. Patients are treated with 1.8 to 2 Gy fractions, five days a week without a planned treatment break.

The field arrangement at Fox Chase Cancer Center consists of four fields (AP, PA, right, and left lateral). In our opinion this arrangement is the most practical and allows for straightforward portal verification.

TREATMENT VERIFICATION

Real time portal imaging is utilized in all patients with prostate cancer treated at Fox Chase Cancer Center. A digital image archiving system presently in place allows for portal verification from a physician workstation connected to the local area network (Figure 2). The increased flexibility of this system should improve both the accuracy and efficiency of treatment delivery.

CASE ILLUSTRATIONS

Patient #1
A 67-year-old man with a history of an elevated PSA (7.6 ng/mL) on a routine screening test. Digital rectal examination revealed a 1 cm nodule in the left mid-gland. Ultrasound-guided fine-needle biopsy was performed and a Gleason grade 5 adenocarcinoma was diagnosed. Bone scan was negative and MRI scan demonstrated an abnormality in the left side of the

Table 3. Published Tolerance Doses of Critical Structures in Patients Receiving External Beam Radiation Therapy for Prostate Cancer[1]

Organ	Volume of Organ	TD 5/5 (cGy)	TD 5/50 (cGy)
Rectum	Complete	6000	8000
Bladder	Complete	6500	8000
	Two-thirds	8000	8500
Small bowel	Complete	4000	5500
	One-third	5000	6000

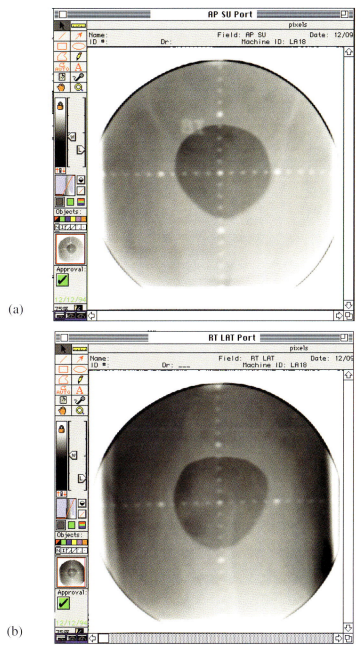

(a)

(b)

Figure 2. On-line portal image as displayed on physician workstation. (a) anterior (b) lateral.

gland with no evidence of extracapsular extension. The patient was staged T2a and external beam radiation therapy was recommended. CT simulation was performed and the DRR for this patient is shown in Figure 3. In this case the GTV and CTV were defined as the entire prostate gland and the PTV was defined as an additional 1 cm beyond the CTV. The treatment plan was to deliver 68 Gy in 34 fractions to the PTV (enclosed within the 95% isodose volume) using an isocentric, four-field technique.

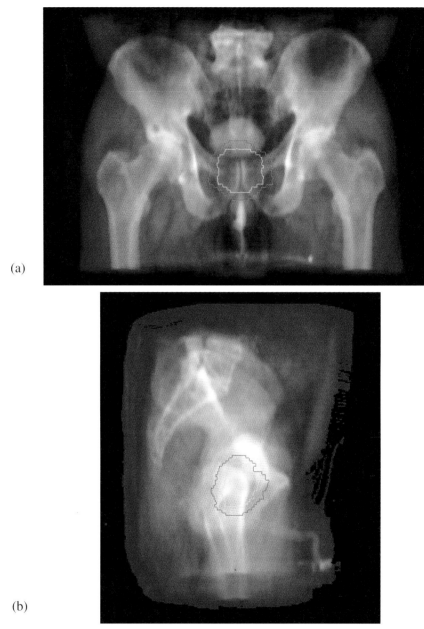

Figure 3. Digitally reconstructed radiograph of patient #1. (a) Anteroposterior view with CTV and PTV outlined. (b) Lateral view with CTV and PTV outlined.

Patient #2

A 71-year-old man with a history of urinary frequency was found to have an enlarged prostate during a yearly physical. Serum PSA was elevated at 17.5 ng/mL and needle biopsy of the prostate gland was performed. Pathologic review of the specimen demonstrated Gleason score 8 adenocarcinoma of the prostate. Bone scan was negative and MRI scan suggested left-sided seminal vesicle involvement. Digital rectal examination revealed left-sided extracapsular penetration. The patient was staged T3c and external beam radiotherapy combined with neoadjuvant androgen deprivation were recommended. CT simulation was performed following two months of hormonal treatment and a DRR was constructed as shown in Figure 4. In this case the GTV included the prostate and left seminal vesicle. The CTV1 included the prostate, both seminal vesicles and the pelvic lymph nodes, the CTV2 included the prostate and bilateral seminal vesicles, CTV3 included the prostate and left seminal vesicle. The PTV was defined as 1 cm beyond the CTV. The treatment plan is to deliver 45 Gy in 28 fractions to PTV1, an additional 12 Gy in six fraction to PTV2 (total dose 57 Gy to PTV2) and a further 14 Gy in seven fractions to PTV3 (total dose 71 Gy to PTV3). Four-field technique is planned throughout. A portal image of a similar patient displayed on a physician workstation is shown in Figure 5.

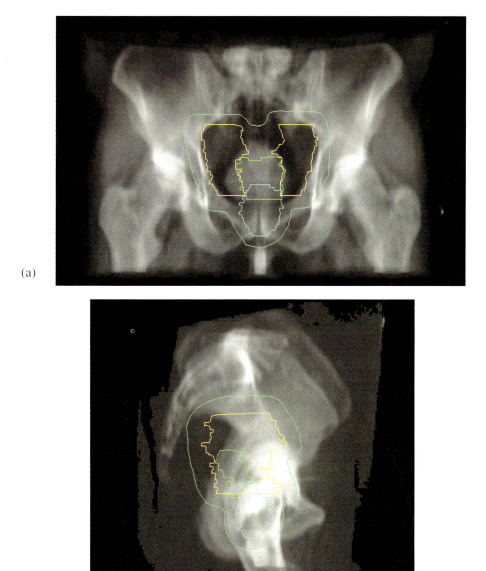

(a)

(b)

Figure 4. Digitally reconstructed radiograph of patient #2. (a) Anteroposterior view with CTV1 and PTV1 outlined. (b) Lateral view with CTV1 and PTV1 outlined.

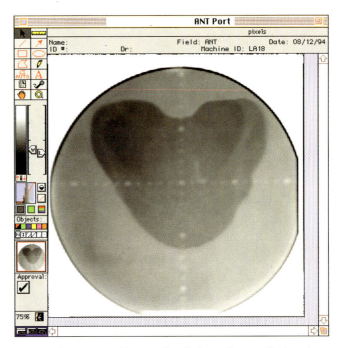

Figure 5. On-line portal image of a similar patient as displayed at a physician workstation.

REFERENCES

1. Hanks GE, Lee WR, Schultheiss TE. Clinical and biochemical evidence of control of prostate cancer at five years after external beam irradiation. J Urol 1995;154:456-459.

2. McNeal JE, Redwine EA, Freiha FS, Stamey TA. Zonal distribution of prostatic adenocarcinoma. Am J Surg Pathol 1988;12:897-906.

3. McNeal JE, Villers AA, Redwine EA, Freiha FS, Stamey TA. Capsular penetration in prostate cancer. Am J Surg Pathol 1990;14:240-247.

4. Partin AW, Yoo J, Carter HB, et al. The use of prostate specific antigen, clinical stage and Gleason score to predict pathologic stage in men with localized prostate cancer. J Urol 1993; 150:10-14.

5. Emami B, Lyman J, Brown A, et al. Tolerance of normal tissue to therapeutic irradiation. Int J Radiat Oncol Biol Phys 1991;21:109-122.

6. Schultheiss TE, Hanks GE, et al. Incidence of and factors related to late complications in conformal and conventional radiation treatment of cancer of the prostate. Int J Radiat Oncol Biol Phys 1995;32:643-649.

Chapter 14

RECTAL CANCER

Rachelle M. Lanciano, MD and Lawrence R. Coia, MD

INTRODUCTION

Rectal cancer is the second most common gastrointestinal malignancy with 43,000 estimated new cases in the United States in 1993.[1] Local recurrence is a common problem following curative surgery alone ranging from 25% to 70% for T3/T4 (penetration through the muscularis propria or invasion of surrounding structures) or N1/2 disease (1 to 3 or greater than 4 perirectal lymph nodes).[2,3,4] With postoperative chemoradiation, the pelvic recurrence rate for the same group of patients is approximately 10%.[5,6,8] The Gastrointestinal Tumor Study Group (GITSG) and the North Central Cancer Treatment Group have proven the superiority of concurrent chemoradiation over surgery alone or surgery plus single modality adjuvant therapy with improved local pelvic control and survival for stage II and III rectal cancer.[5,6] In 1990, a National Institutes of Health Consensus Development Conference recommended combined chemoradiation as standard practice for this group of patients.[7] A recent GITSG study demonstrated a significant decrease in the overall rate of tumor relapse and distant metastases with protracted infusion compared to bolus Fluorouracil during radiation following curative resection.[8] The current intergroup postoperative adjuvant study is exploring the use of continuous infusion Fluorouracil prior to and following radiation as well as the use of Leucovorin and Levamisole with bolus Fluorouracil.[9] In addition, an intergroup study is comparing preoperative chemoradiation to postoperative chemoradiation for clinical stage T3/T4 and N1/N2 patients to evaluate if sphincter preservation and function, quality of life, and tolerance of treatment may be improved with preoperative chemoradiation which may translate into improvements in overall survival.[10]

Despite the relatively high local control rates and low morbidity of combined modality treatment for rectal cancer, there is room for improvement. Although local failure is only 10% for stage II and III cancers managed with surgery and postoperative chemoradiation, it is approximately 30% for patients with tethered, fixed, or T4 lesions. Furthermore, the adjuvant use of 55 Gy has been associated with a higher local control rate for such advanced lesions compared to 45 Gy.[11] In addition, the toxicity of combined chemoradiation is substantial with approximately 40% of patients experiencing grade 3 diarrhea and 35% and 15% of patients, respectively, never completing all the planned chemotherapy cycles due to toxicity or patient refusal.[6] CT simulation may help better define both the target volume and dose limiting structures, so that radiation dose can be escalated to the target volume and the dose to normal structures of importance such as small bowel can be minimized.

ANATOMY

The rectum is defined anatomically as extending from the third sacral segment to slightly below the tip of the coccyx with a total length of 12 cm.[12] In reality, trials involving patients with rectal cancer generally require the lower edge of the tumor mass to be below the peritoneal reflection, or a portion of the tumor to be retroperitoneal, or the lower margin of the tumor to be 12 cm or less from the anal verge by proctoscopic examination.[9,10]

The rectum deviates from anterior and left sided to posterior and right sided from caudal to cephalic position. There are three permanent transverse folds (Houston's valves) which project into the lumen with the largest and lowest adjacent to the bladder. Unlike the colon, the rectum has a complete outer longitudinal muscle coat and the caudal end is dilated to form the rectal ampulla (Figure 1).

The upper two-thirds of the rectum has some peritoneum located anterior and lateral in the most cephalad portion and only lateral in the middle portion. The lower third has no peritoneum. Posterior to the rectum are the superior rectal vessels, piriformis muscle, sacral nerve plexus and the fascia covering the sacrum, coccyx and levator ani muscle. Anteriorly in the male, the rectum is separated by intestine in the rectovesical fossa from the fundus of the bladder and by the rectovesical septum from the seminal vesicles and prostate. Anteriorly in the female, the rectum is separated by intestine in the rectouterine fossa from the uterus and by the rectovaginal septum from the posterior wall of the vagina (Figure 2).

PATTERNS OF SPREAD

Lymph node drainage from the rectum is through the perirectal nodes which lie on the rectal muscles and sigmoid mesocolon. If there is involvement of the anal canal, the lymph drainage can accompany the middle rectal artery and end in the internal iliac nodes and at or below the pectinate line to the inguinal lymph nodes[12] (Figure 3).

Detailed information concerning patterns of failure after curative surgery have been reported from the University of Minnesota reoperation series of patients either at high risk of

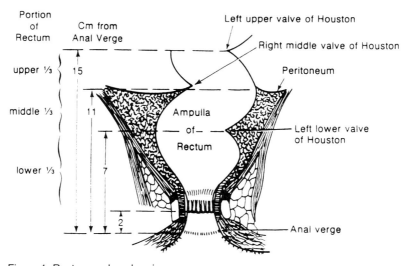

Figure 1. Rectum and anal region.

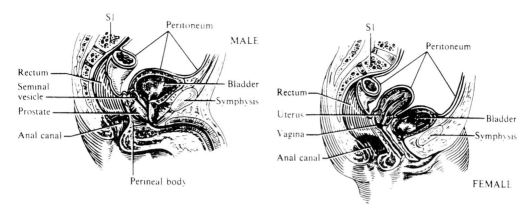

Figure 2. Male/female rectal peritoneal relations.[19]

recurrence or symptomatic. Locoregional recurrence in the pelvis or paraaortic region was the sole pattern of failure in 48% of patients and occurred as a component of failure in 92%.[13] Idealized radiation fields have been generated to encompass these patterns of failure.[14] With adjuvant chemoradiation, the predominant pattern of failure is distant (30% to 40%), most commonly in the liver, lung, distant lymph nodes, and peritoneum. Local failure with combined chemoradiation ranges from 6% to 13%.[5,6,8]

DIAGNOSTIC STUDIES

Physical examination remains the most important pretreatment evaluation and should include mobility of the tumor (mobile, tethered or fixed), the level above the anal verge or pectinate line, size and percentage of circumferential involvement. Proctoscopy and barium enema or colonoscopy complements the digital examination, provides tissue diagnosis, and

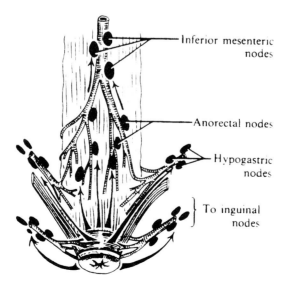

Figure 3. Rectal lymph drainage.[19]

rules out synchronous large bowel pathology. Preoperative computerized tomography aids in the diagnosis of pelvic adenopathy when lymph node size is ≥ 1.5 cm, assesses distant spread to the liver, helps determine if a free space exists between the tumor and surrounding structures, and documents the level of the tumor mass in relation to bony landmarks. Pelvic MRI either with a Helmholtz or endorectal coil can improve sensitivity and resolution, compared with that achieved with a body coil alone for detection of transmural disease and enlarged perirectal lymph nodes. Endorectal ultrasound has improved the accuracy in evaluation of transmural extension and enlarged perirectal lymph nodes preoperatively.[15] At the Fox Chase Cancer Center we are studying the accuracy of both modalities (Helmholtz MRI and endorectal ultrasound) in detecting transmural extension/contiguous organ involvement and lymph node metastases. Additional tests required for preoperative evaluation include routine blood studies, carcinoembryonic antigen, and chest x-ray. The staging system for rectal cancer is shown in Tables 1 and 2.

IMMOBILIZATION

Patients undergo CT simulation in the prone position with their arms overhead. The prone position is generally used since small bowel may shift superiorly (out of the pelvis) with anterior compression by the tabletop and the perineum is more easily visualized for accurate placement of markers and bolus material. The supine position is used if the external iliac nodes are to be included or occasionally in male patients if scrotal shielding is indicated. For immobilization, a custom Alpha Cradle from the iliac crest to the knees is made. The patient's face rests in a face mask and a triangle sponge is placed under the ankles for support.

LOCALIZATION

Definition of Gross Tumor Volume

The gross tumor volume (GTV) represents the tumor in the rectum seen on CT scan if treatment is given preoperatively (usually for T3/T4 or N1/N2). In the postoperative setting, the GTV should be reconstructed from the preoperative CT scan and/or barium enema.

Definition of Clinical Target Volume

The initial clinical target volume (CTV1) represents the GTV plus 5.0 proximally and distally for microscopic extension along the bowel wall, the entire mesorectum circumferentially to the lateral sidewall (obturator muscle), anteriorly to the vaginal marker in women, and the bladder/prostate/seminal vesicles in men, posteriorly to the anterior sacrum (or for patients status post APR to the perineal scar). The CTV1 also encompasses the draining lymphatics and includes the perirectal nodes and the internal iliac nodes. If the tumor has clinically or pathologically involved the bladder, prostate, uterus or vagina, the external iliac lymph node system is at risk and included in the CTV1.

The Planning Target Volume

The planning target volume (PTV1) represents the CTV1 plus a margin necessary for day-

Table 1. Definition of TNM

Primary Tumor (T)
TX Primary tumor cannot be assessed.
T0 No evidence of primary tumor.
Tis Carcinoma in situ: intraepithelial or invasion of the lamina propria.*
T1 Tumor invades the submucosa.
T2 Tumor invades the muscularis propria.
T3 Tumor invades through the muscularis propria into the subserosa or into nonperitonealized pericolic or perirectal tissues.
T4 Tumor directly invades other organs or structures and/or perforates the visceral peritoneum.**

*Tis includes cancer cells confined within the glandular basement membrane (intraepithelial) or lamina propria (intramucosal) with no extension through the muscularis mucosae into the submucosa.
**Direct invasion of other organs or structures includes invasion of other segments of colorectum by way of serosa; for example, invasion of the sigmoid colon by a carcinoma of the cecum.

Regional Lymph Nodes (N)
NX Regional lymph nodes cannot be assessed.
N0 No regional lymph node metastasis.
N1 Metastasis in one to three pericolic or perirectal lymph nodes.
N2 Metastasis in four or more pericolic or perirectal lymph nodes.
N3 Metastasis in any lymph node along the course of a named vascular trunk and/or metastasis to apical nodes(s) (when marked by the surgeon).

Distant Metastasis (M)
MX Presence of distant metastasis cannot be assessed.
M0 No distant metastasis.
M1 Distant metastasis.

to-day set-up variability. The daily set-up variability is 1 cm for the pelvis which represents two standard deviations around the mean.[17]

The CTV2 represents the GTV1 plus a minimum of 2 cm margin both proximally and distally. The PTV2 represents the CTV2 plus 1 cm.

DEFINITION OF CRITICAL STRUCTURES

The critical structures of concern in the pelvis include the bladder, rectum, femoral head, and small bowel which have maximum tolerable doses of 6500 cGy, 6000 cGy, 5200 cGy and 5000 cGy, respectively.[18] Since typical doses delivered for preoperative or postoperative radiation are in the range of 5000 to 5400 cGy for rectal cancer, the only critical structure above the TD5/5 is the small bowel which needs to be blocked after 5000 cGy has been delivered. It tends to be easier to shield small bowel when patients are treated preoperatively since the rectum in situ moves small bowel anteriorly and superiorly, thereby allowing significant shield-

Table 2. Stage Grouping

AJCC/UICC				Dukes
Stage 0	Tis	N0	M0	...
Stage I	T1	N0	M0	A
	T2	N0	M0	
Stage II	T3	N0	M0	B*
	T4	N0	M0	
Stage III	Any T	N1	M0	C
	Any T	N2	M0	
	Any T	N3	M0	
Stage IV	Any T	Any N	M1	...

*Dukes B is a composite of better (T3, N0, M0) and worse (T4, N0, M0) prognostic groups, as is Dukes C (Any T, N1, M0 and Any T, N2 N3, M0).

ing by lateral field blocking. Additionally, patients are treated with a full bladder to reduce the volume of small bowel in the pelvic fields.

CT LOCALIZATION OPTIMIZATION AND VIRTUAL SIMULATION

Treatment localization and CT simulation procedures are performed by the therapist (technologist) using the guidelines outlined in Chapter 3. For rectum specifically, the isocenter is defined in the treatment localization mode using the CTV1 outlined by the physician. The

Table 3. Definition of Target Volumes

GTV	Preoperative	Gross disease on CT scan
	Postoperative	Reconstruction of gross disease from preoperative CT and/or barium enema, includes anastomosis.
CTV1	Preoperative	Internal iliacs, perirectal lymphatics, + external iliacs*, plus GTV + 5 cm proximal and distal, laterally to pelvic sidewall, anteriorly to bladder/prostate (men) or vaginal marker (women), posterior to sacrum.
	Postoperative	Same as preoperative, except: Perineal scar is included posteriorly if s/p APR.
PTV1	=	CTV1 + 1 cm
CTV2	=	GTV + 2 cm
PTV2	=	CTV2 + 1 cm
PTV1		45 Gy
PTV2		50.4-54.0 Gy

*External iliacs included if bladder, vagina, or prostate involved.

physician encompasses lymphatic drainage to be included in the CTV1 by identifying appropriate major vessels (e.g., internal iliacs, common iliacs ± external iliacs). Beam arrangement optimization (part of the virtual simulation process) is then carried out by the dosimetrist and/or physician. DRRs of the planned beam fields with target, critical structures, and selected normal structures outlined are obtained and conformal beam blocks are drawn using the treatment planning guidelines mentioned earlier.

Patients are given 450 cc of ready to use barium sulfate (1.2%) suspension for computed tomography (Readi-CAT, E-Z-M Inc). The patient is instructed to drink the entire bottle one hour before casting so that contrast is optimized in the small bowel at the time of simulation (approximately one and one-half hours following ingestion).

Campostrini has described the technique for visualization of pelvic structures using conventional simulation and we have adapted this technique for CT simulation.[16] A 22 Fr Foley catheter is placed per rectum with wire solder inside the catheter for visualization on x-ray. A scout film is taken to assure the Foley catheter is in the appropriate position above the pubic symphysis. The solder wire is then removed and 1 cc of Renograffin 30 mixed with 8 cc of sterile water in a 10 cc syringe is used to inflate the balloon. The Foley catheter is pulled inferior so that the Foley balloon is tightly against the anal canal lying in the rectal ampulla. To visualize the rectum, 60 cc of Readi-CAT is injected into the Foley catheter which is subsequently clamped to prevent outflow of contrast from the catheter. A CT-compatible anal marker is applied to the anal verge with attached metallic chains used to define the perineum/intergluteal fold and anal verge. With this technique the anal verge, anal canal, and rectal ampulla are easily defined on CT images. In females, a large Q-tip is soaked in Readi-CAT, placed in a glove and inserted into the vagina at the apex.

Each patient is scanned from the top of the iliac crest to 1 cm below the ischial tuberosity or below perineal marker. The patient is imaged with overlapping CT slices that are 5 mm thick at 3 mm intervals.

TREATMENT PLANNING

Dose Prescription for Target Volume

The following guidelines for radiation dose prescription were formulated on the basis of delivery of concurrent chemotherapy (infusional 5-Fluorouracil) and radiation therapy. The PTV1 is typically treated with a three- or four-field plan (PA ± AP plus laterals) with shaped blocks. The PTV1 is encompassed by the 95% isodose treated to 45 Gy. The PTV2 receives an additional 5.4 Gy at a minimum or an additional 9.0 Gy if small bowel can be entirely excluded from the PTV2. The total dose to the PTV2 is, therefore, 50.4 to 54.0 Gy.

TREATMENT VERIFICATION

Digitally reconstructed radiographs (DRR) alone *do* provide enough information for treatment verification since the pelvic bones such as the pubic symphysis, iliac crest, acetabulum and sacroiliac joint are well delininated on a DRR. Contrast in small and large intestine, vaginal and anal markers, and perineal/intergluteal fold chains may or may not be seen on the DRR. Therefore, if indicated, these structures may need to be outlined on the CT simulator for visualization on DRR.

CASE ILLUSTRATIONS

Case #1

FP is a 50-year-old white male who presented with intermittent rectal bleeding (thought to be hemorrhoids), constipation for 6 months, 34 pound weight loss and pelvic/anal pain. Rectal examination revealed a circumferential multilobulated friable mass 10 cm from the anal verge with almost complete obstruction. Biopsy revealed moderately differentiated adenocarcinoma. Helmholtz MRI of the pelvis revealed an extensive circumferential high rectal mass extending to the rectosigmoid junction. There was extension into the perirectal fat, and at the level of the rectosigmoid junction the tumor extended into the presacral space and abutted the sacrum. There was evidence of perirectal adenopathy. CT scan of the chest and abdomen revealed no evidence of metastatic disease. Patient was enrolled on the FCCC phase I trial of radiation dose escalation with twice daily pelvic radiation and concurrent infusional 5-Fluorouracil and Leucovorin. As depicted in Figure 4, the CTV1 and PTV1 are quite large due to the extensiveness of the primary, probable contiguous spread to the bladder, and inclusion of the external iliac vessels. A four-field box technique was used to encompass this volume. Figure 5 shows posterior and lateral DRRs with CTV1 and PTV1 depicted.

Case #2

RL is a 51-year-old white female with a history of a left colectomy for a T3N1 adenocarcinoma of the sigmoid colon in December 1993. Postoperatively the patient received 6 months of 5-Fluorouracil and Levamisole. On routine surveillance colonoscopy a recurrence was noted at the anastomosis which was confirmed by biopsy. CT scan of the abdomen/pelvis revealed thickening of the distal sigmoid at the level of the anastomosis but no adenopathy or liver metastases. The patient is being treated preoperatively on the Radiation Therapy Oncology Group (RTOG) study 93-03 for recurrent rectal cancer with protracted venous infusion of 5-Fluorouracil 300 mg/m^2/day Monday through Friday during the entire course of pelvic radiation to a dose of 5040 cGy, followed by surgical resection plus intraoperative radiation therapy. As depicted in Figure 6, the CTV1 encompasses the target volume, which includes the rectosigmoid junction and perirectal nodes with 5 cm above and below the anastomosis. Note the vaginal marker to define the anterior extent of the target. Figure 7 shows posterior and lateral DRRs with CTV1 and PTV1 depicted.

RECTAL CANCER

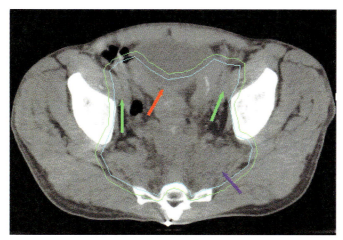

Figure 4a. Note contiguous involvement of bladder (red arrow) and extension of tumor into the sciatic notch causing pain (blue arrow). CTV encompasses external iliac vessels (green arrow).

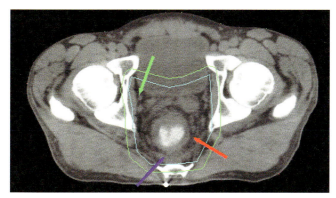

Figure 4b. Note perirectal adenopathy (red arrow), presacral thickening (blue arrow), and anterior extension of CTV to encompass entire perirectal fat and adenopathy (green arrow).

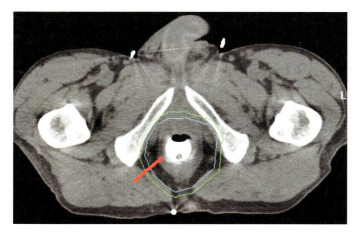

Figure 4c. Inferior aspect of field with Foley balloon in rectal ampulla (red arrow).

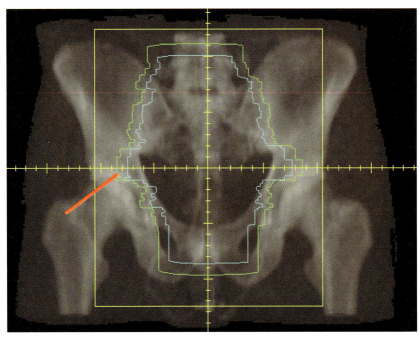

Figure 5a. Locally advanced rectal cancer — anterior DRR. Note lateral extension of external iliac vessels (red arrow). CTV1: Blue. PTV1: Green.

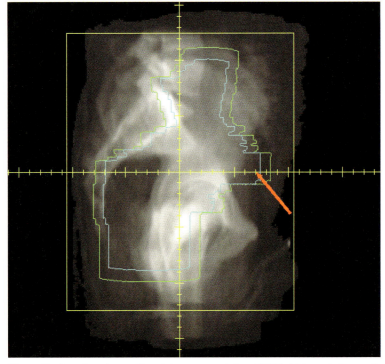

Figure 5b. Locally advanced rectal cancer — lateral DRR. Note anterior position of external iliac vessels (red arrow).

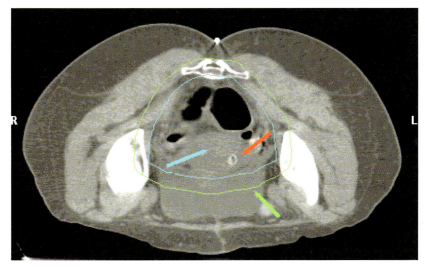

Figure 6a. CTV1 encompasses from the sacrum posteriorly to the vagina anteriorly. Note vaginal marker (red arrow), uterus (blue arrow), and bladder (green arrow). CTV1: Blue. PTV1: Green.

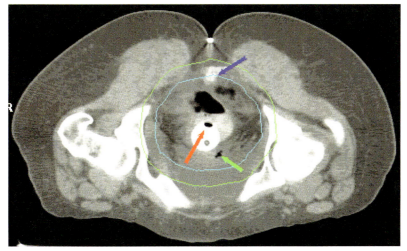

Figure 6b. CTV1 encompasses from the anterior coccyx (blue arrow) to the anterior vaginal marker (green arrow). Note Foley balloon in rectal ampulla (red arrow).

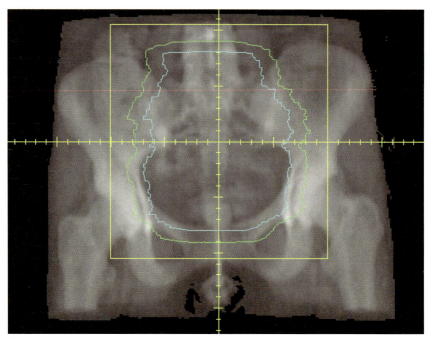

Figure 7a. Anterior DRR. CTV1: Blue. PTV1: Green.

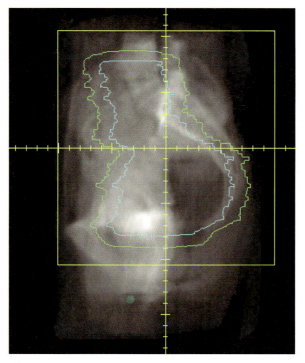

Figure 7b. Lateral DRR.

REFERENCES

1. Boring CC, Squires TS, Tong T. Cancer statistics, 1993. CA Cancer J Clin 1993;43:7-26.

2. Rich T, Gunderson LL, Lew R, Galdibini JJ, Cohen AM, Donaldson G. Patterns of recurrence of rectal cancer after potentially curative surgery. Cancer 1983;52:1317-1329.

3. Gilbert SG. Symptomatic local tumor failure following abdomino-perineal resection. Int J Radiat Oncol Biol Phys 1978;4:801.

4. Walz BJ, Green MR, Lindstrom ER, Butcher HR. Anatomical prognostic factors after abdominal perineal resection. Int J Radiat Oncol Biol Phys 1981;7:477-484.

5. Douglass HO, Moertel CG, Mayer RJ, Thomas PRM, Lindblad AS, Mittleman A, Stablein DM, Bruckner HW (Letter to the Editor). Survival after postoperative combination treatment of rectal cancer. N Engl J Med 1986;315:1294-1295.

6. Krook JE, Moertel CG, Gunderson LL, Wieand HS, Collins RT, Beart RW, Kubista TP, Poon MA, Meyers WC, Mailliard JA, Twito DI, Morton RF, Veeder MH, Witzig TE, Cha S, Vidyarthi SC. Effective surgical adjuvant therapy for high-risk rectal carcinoma. N Engl J Med 1991;324:709-715.

7. NIH Consensus Conference. Adjuvant therapy for patients with colon and rectal cancer. JAMA 1990;264:1444-1450.

8. O'Connell MJ, Martenson JA, Wieand HS, Krook JE, MacDonald JS, Haller DG, Mayer RJ, Gunderson LL, Rich TA. Improving adjuvant therapy for rectal cancer by combining protracted-infusion Fluorouracil with radiation therapy after curative surgery. N Engl J Med 1994;331:502-507.

9. Radiation Therapy Oncology Group. Postoperative evaluation of 5-FU by bolus injection vs 5-FU by prolonged venous infusion prior to and following combined prolonged venous infusion plus pelvic XRT vs bolus 5-FU plus Leucovorin plus Levamisole prior to and following combined pelvic XRT plus bolus 5-FU plus Leucovorin in patients with rectal cancer, Phase III. RTOG Protocol 94-03, 1994.

10. Radiation Therapy Oncology Group. Phase III intergroup randomized trial of pre-operative vs post-operative combined modality therapy for resectable rectal cancer. RTOG Protocol 94-01, 1994.

11. Ahmad NR, Marks G, Mohiuddin M. High-dose preoperative radiation for cancer of the rectum: Impact of radiation dose on patterns of failure and survival. Int J Radiat Oncol Biol Phys 1993;27:773-778.

12. Pansky B. Rectum and anal canal. In Pansky B: Review of Gross Anatomy, 4th edition. New York, New York: Macmillan, 1979, pp 372-375.

13. Gunderson LL, Sosin H. Areas of failure found at reoperation (second or symptomatic look) following "curative surgery" for adenocarcinoma of the rectum: Clinicopathologic correlation and implications for adjuvant therapy. Cancer 1974;34:1278-1292.

14. Gunderson LL, Russell AH, Llewellyn HJ, Doppke KP, Tepper JE. Treatment planning for colorectal cancer: Radiation and surgical techniques and value of small bowel films. Int J Radiat Oncol Biol Phys 1985;11:1379-1393.

15. de Lange EE. Staging rectal carcinoma with endorectal imaging: how much detail do we really need? Radiology 1994;190:633-635.

16. Campostrini F, Garusi G, Donati E. A practical technique for conformational simulation in radiation therapy of pelvic tumors. Int J Radiat Oncol Biol Phys, 1995. In press.

17. Hunt M, Schultheiss T, Desobry G, Hakki M, Hanks GE. An evaluation of setup uncertainties for patients treated to pelvic sites. Int J Radiat Oncol Biol Phys, 1995. In press.

18. Emami B, Lyman J, Brown A, Coia L, Goitein M, Munzenrider JE, Shank B, Solin LJ, Wesson M. Tolerance of normal tissue to therapeutic irradiation. Int J Radiat Oncol Biol Phys 1991;21:109-122.

19. Pansky B, House EL. Review of Gross Anatomy, 2nd edition. London: McMillan Co., 1969, pp 359-360.

Chapter 15

BREAST CANCER

Eileen F. McGarvey, MD

INTRODUCTION

Breast cancer now strikes one in nine women in this country. For many women with stage I and II disease, breast preservation treatment consisting of lumpectomy, axillary lymph node dissection, and breast irradiation is an acceptable treatment alternative to modified radical mastectomy. Radiation therapy treatment planning for the breast and regional lymphatics is standardized within our department using conventional simulation. We do not use CT simulation for planning of the breast and regional nodes. This is because of the ease with which accurately devised treatment plans are obtained with conventional simulation, the difficulty in delineating breast and regional nodes on CT, the poor quality of DRRs at this site, and the increased cost and labor intensity required using CT treatment planning. We have found CT treatment planning to be an extremely useful tool to use in delineating the boost field, particularly when the tumor bed is deep-seated, when there are no clips at the surgical site, when the surgical scar is not situated over the area of the breast where the mammographic abnormality or palpable disease was noted, and when it is difficult to correlate the mammographic site of disease with the area of disease within the breast with the patient in the treatment position. In these cases CT has been helpful in identifying the changes within the breast consistent with the biopsy site, the depth to the deepest portion of the tumor bed, the treatment angle, and the volume of breast to be included within the boost field. For patients with tumors 5 cm or less in depth, the boost field is treated with an electron beam. For patients with tumors greater than 5 cm in depth, a three-field photon beam boost plan is generated from the information gathered using CT boost treatment planning.

ANATOMY

The breast is situated on the anterior and anterior lateral chest wall overlying the pectoral major and the serratus anterior muscles. It extends from the second rib to the sixth costal cartilage and from the sternal edge to the midaxillary line. The tail of the breast extends into the axilla along the inferior edge of the pectoral major muscle. The breast is separated from the underlying muscles by a fascial layer to which it is loosely attached, and it is tightly bound to the overlying skin by a group of ligaments (Cooper's ligaments), which serve to support the breast.[1]

The *axilla* is the primary nodal drainage site of the breast. The axilla is bounded anteriorly by the clavicle and pectoral muscles, posteriorly by the scapula and subscapularis muscles, medially by the ribs, intercostal muscles and serratus anterior muscles, laterally by the medial aspect of the proximal humerus, superiorly by the axillary apex (supraclavicular area) and inferiorly by the axillary skin. The axilla is divided into three levels. Those nodes within the axilla located lateral to the lateral edge of the pectoralis minor muscle are within level I. Level II nodes are deep to the pectoralis minor muscle and level III nodes are medial to the medial edge of the pectoralis minor muscle. The *supraclavicular nodal area* is a distinct anatomic region bounded by the clavicle, the scapula, the first rib, and the coracoid process.[1] The *internal mammary nodes* are situated within 3 cm of the lateral margin of the sternum in the interspaces between the costal cartilages with most of the nodes between the first three interspaces.[2]

PATTERNS OF SPREAD

Within the breast, malignant cells can spread by direct infiltration into the breast parenchyma, along the ductal system, or through the lymphatics of the breast. In advanced disease, the skin or pectoralis fascia and muscle may be involved. The axillary area is the most common site of regional spread for all sites of disease within the breast. Metastatic disease can also be seen in internal mammary nodes and in the supraclavicular nodes.

DIAGNOSTIC STUDIES

The mammogram is used to evaluate the breast. If microcalcifications are noted prior to biopsy, a post-biopsy mammogram is obtained to confirm that all microcalcifications have been removed prior to offering breast irradiation. Chest CT is useful in evaluating regional nodal areas and the chest when clinically indicated such as suspicion of fixed axillary nodes or chest wall invasion.

PATIENT POSITIONING

Our preference is to treat the breast tangents and the breast boost field with the patient in the same position whenever possible. The patient is supine on a breast board in an Aquaplast cast with arms raised and hands behind head. Because of the fixed diameter of the CT aperture (70 cm), and because of the wide variation in anatomy and size of patients, not all patients can pass through the CT scanner on a breast board. In these cases the breast board is omitted. For far lateral lesions patients may be positioned in the lateral decubitus position with the ipsilateral arm raised with the hand behind the head.

IMMOBILIZATION

An Aquaplast cast extending from the mid neck to iliac crests is used.

LOCALIZATION

Definition of Gross Tumor Volume

In breast CT treatment planning, the term gross tumor volume (GTV) applies in the situation where breast grossly involved with tumor exists or where gross nodal involvement is identified. For patients who have had a lumpectomy and an axillary node dissection, this term is not applicable.

Definition of Clinical Target Volume

The initial clinical target volume (CTV1) is the entire breast alone plus the underlying chest wall. CTV1 may also include regional nodes as dictated by the clinical situation. The clinical target volume (CTV2) applies to tumor bed defined by clips placed at the time of lumpectomy or by correlation of mammographic findings with surgical distortion at the biopsy site within the breast seen on planning CT scan. The margin used to define the CTV2 for the breast boost depends on the clinical situation, i.e., surgical margin status, presence of lymphatic or vascular invasion, presence of ductal carcinoma in situ, depth of the tumor bed, etc. When clips are not placed post-lumpectomy and the exact location of the lesion cannot be readily discerned, the CTV2 extends from scar to chest wall.

Definition of Planning Target Volume

Planning target volume (PTV) is ordinarily defined as CTV + 1 cm to account for motion and set up error. When using electrons to treat the PTV2 we must take into account the narrowed dose distribution at depth. Therefore, the margin around the CTV will be greater than 1 cm to adequately treat the entire tumor bed.

DEFINITION OF CRITICAL STRUCTURES

Within the fields used to treat breast and regional lymph nodes are a number of critical structures at potential risk for the development of treatment-related complications. Among these are lung, heart, contralateral breast, muscle, ribs, clavicle, brachial plexus, ipsilateral humeral head, and skin.[3] The risks of complications to any of these sites increases as the volume of and dose to these structures increases. The two most critical structures are the heart and lung. These organs are easily identified on CT and their volume within the treatment field(s) can be readily determined. The Toxicology Working Group has defined the TD 5/5 and TD 50/5 for these organs as a function of dose and area treated (see Table 1).[4]

CT SCAN LOCALIZATION OPTIMIZATION AND VIRTUAL SIMULATION PROCESS

As noted in the introduction, we do not use CT simulation for planning treatment to the breast or regional nodes. However, we do use CT simulation in treatment of the cone-down boost, i.e., treatment to the tumor bed (CTV2). In some cases the biopsy site is well delineat-

Table 1.[4]

Organ	Injury	TD 5/5	TD 50/5	Whole or Partial Organ
Lung	Acute & chronic pneumonitis	3000 1500	3500 2500	100 cm^2 Whole lung
Heart	Pericarditis and pancarditis	4500 7000	5500 8000	60% 25%

ed on CT and may or may not be directly beneath the biopsy scar. In other cases the biopsy site is difficult to ascertain, and in others the tumor bed is too deep for an electron boost. In these situations CT planning of the breast boost has been very useful. The following examples will serve to illustrate our use of CT for breast boost treatment planning. Treatment localization and CT simulation procedures are performed by the therapist (technologist) using the general guidelines outlined in Chapter 3. Beam arrangement optimization (part of the virtual simulation process) is then carried out by the dosimetrists and/or physician.

CONCLUSION

CT is very useful for planning the breast boost. Areas of surgical distortion and/or clips identifying the tumor bed, the depth of the tumor bed, and the treatment gantry angle can all be defined readily using this technique.

CASE ILLUSTRATIONS

Patient #1

A 35-year-old noted a mass in the high upper central to upper outer quadrant of her left breast. Mammogram showed no suspicious calcifications or densities. Magnification views of the area of palpable abnormality were unremarkable. Ultrasound showed a cyst at the 12 to 1 o'clock position. Fine needle aspiration was remarkable for malignant cells. Lumpectomy yielded a 2 cm poorly differentiated infiltrating ductal carcinoma. Clips were not left at the surgical site. A curvilinear lumpectomy scar, 5 cm in length, was located 2.5 cm superior to the areolar complex at the 12 o'clock position. By the patient's report, the palpable mass was situated several centimeters superior to the scar, in an area of obvious surgical volume loss in the upper outer quadrant.

The patient received whole breast radiation therapy for a total dose of 4600 cGy in 200 cGy daily fractions followed by a boost to the tumor bed for an additional 2000 cGy resulting in a total tumor bed dose of 6600 cGy. The tumor bed was identified using CT treatment planning. During CT, wires were placed on the lumpectomy scar and over an area of volume loss at the pre-surgical site of palpable disease (Figure 1). CT slices were taken through the upper half of the breasts. An area of volume loss (the boost site) was identified superior to the lumpectomy scar at the site of initial palpable disease. At that site the depth to the chest wall was determined to be 3 cm (Figure 2). The patient was boosted with a 0° gantry angle using

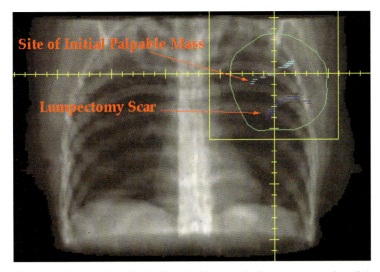

Figure 1. 35-year-old patient with palpable mass in the upper central portion of the left breast. Mammogram showed no suspicious densities or calcifications. A 2 cm infiltrating ductal carcinoma was noted at biopsy. A lumpectomy surgical incision was not placed over the palpable mass and the tumor bed was not marked with clips. At the time of CT simulation a wire was placed over the lumpectomy scar and a second wire was placed over the site of palpable disease. The site of palpable disease actually corresponded to an area of volume loss noted both on physical exam and on CT scan. The circular area encompassing both wires defines the electron boost field.

12 MeV electrons to the 91% isodose line. The circular area in Figure 1 encompassing both wires defines the treated boost field.

Patient #2

A 44-year-old female noted a firm, mobile, marble-sized mass in the upper outer quadrant of the left breast on breast self-examination. Mammogram showed bilateral fibrocystic changes without a dominant density in the area of palpable abnormality. At lumpectomy she was found to have a 2.5 cm high-grade, infiltrating ductal carcinoma. The lumpectomy incision was directly over the palpable mass. Tumor bed clips were not placed at the time of lumpectomy.

The patient received whole breast radiation therapy to 4600 cGy in 200 cGy daily fractions followed by a boost to the CT defined tumor bed to a total dose of 6600 cGy.

At the time of CT treatment planning 3 mm slices were taken through the upper half of the breast. The lumpectomy scar and the areola were marked with wire and the area of tissue distortion beneath the scar, the tumor bed (CTV2), was outlined. A DRR was generated showing the CTV2, the lumpectomy scar and the areolar complex (see Figure 3). The tumor bed (CTV2), the depth of the tumor bed, and the gantry angle were all determined by CT (Figure 4). The deepest aspect of the tumor bed was 5 cm in depth and a gantry angle of 60° was selected for treatment. The CTV2 plus a margin was treated with 20 MeV electrons to the 92% isodose line. The DRR was used for clinical set up.

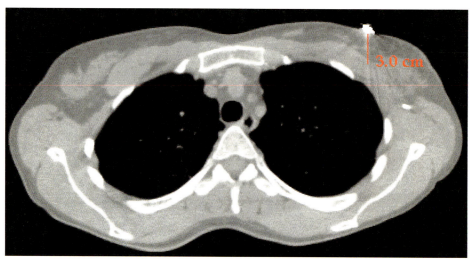

Figure 2. In example #1 an area of surgical volume loss is noted directly under the wire on the skin surface marking the area of presurgical palpable disease. This area of the breast was boosted. Depth to the pectoral muscle was 3 cm, gantry angle 0°.

Patient #3

A 67-year-old female noted a mass in the lower inner quadrant of the left breast on breast self-examination. A mammogram confirmed the presence of a 1.5 cm spiculated density in the lower inner quadrant close to the chest wall. Fine needle aspiration revealed ductal carcinoma. Biopsy revealed a 2.2 cm poorly differentiated infiltrating ductal carcinoma with extensive cribriform intraductal component with negative margins of resection. Clips were placed along the biopsy cavity. The deepest clip was 6.0 cm (Figure 5), beyond the depth treated with our highest energy electron beam. After the patient received 5000 cGy to the whole breast, a boost to the tumor bed was delivered using a three-field photon technique. The tumor bed (CTV2),

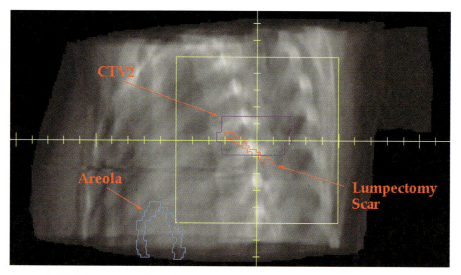

Figure 3. Digital reconstructive radiograph of the CTV2 of patient #2. The areolar complex and the lumpectomy scar were marked with wire and CTV2 defined. A 2 cm margin was placed around CTV2 and cerrobend cut-out made for the electron treatment field. The digital reconstructive radiograph was used for clinical set-up.

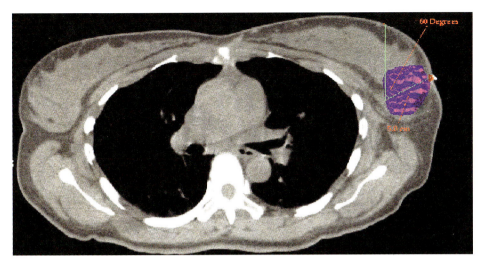

Figure 4. For patient #2 the dark grey area in this CT slice defines the CTV2. As noted, the deepest portion of the tumor bed was 5 cm deep and a gantry angle of 60° was used for treatment.

which was completely encompassed by the clips, was easily identified on CT. After the tumor bed (CTV2) was identified and a plan generated, beam's eye view DRRs of the three fields were generated (Figures 6a, b, and c). These fields were verified and marked on skin during conventional simulation. A 1.5 cm margin was placed around the CTV2 and the volume defined by these fields was treated with 6 MV photons to the 95% isodose line for 7 fractions of 200 cGy each resulting in a total tumor bed dose of 6400 cGy. Figure 7 shows the three-field plan beam arrangement.

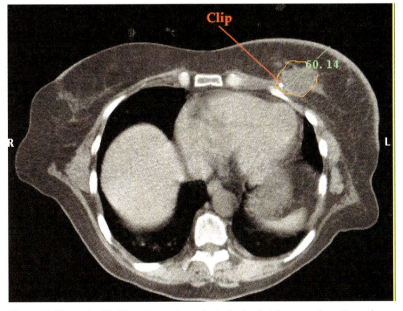

Figure 5. Example #3. The deepest surgical clip is slightly more than 6 cm deep. An area of tissue distortion within the breast, which was completely within the clipped area, was noted at the time of CT and was considered to be the CTV2.

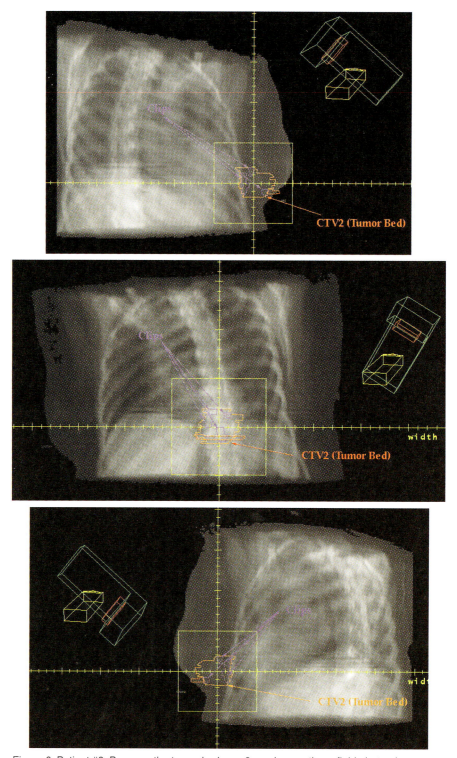

Figure 6. Patient #3. Because the tumor bed was 6 cm deep, a three-field photon boost was planned. These figures show the digital reconstructed radiographs showing the CTV2 in the three different treatment fields. CTV2 and surgical clips are marked. A 1.5 cm margin was placed around the CTV2 in each field. All fields were verified with conventional simulation and tattooed prior to treatment.

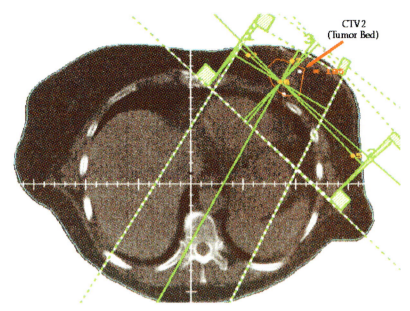

Figure 7. For the patient in example #3. This shows the arrangement of the three fields used for treatment and the CTV2 encompassed by clips.

REFERENCES

1. Moore KL. Clinically Oriented Anatomy. Baltimore, Maryland: Williams & Williams, 1980, pp 680-682, 687.

2. Levin SH, Tapley NV. Technological Basis of Radiation Therapy: Fractional Clinical Applications. Philadelphia, Pennsylvania: Lea & Febiger, 1984, p 229.

3. Solin LJ, Chu JCH, Sontag MR, Brewster L, Cheng E, Doppke K, Drzymala RE, Hunt M, Kuske R, Manolis JM, McCormick B, Munzenrider JE. Three-dimensional photon treatment planning of the intact breast. Int J Radiat Oncol Biol Phys 21:193-203;1991.

4. Toxicology Working Group. Int J Radiat Oncol Biol Phys 5:685;1979.

5. Fraass BA, Lichter AS, McShaw DL, Yanke BR, Diaz RF, Yeakel KS, van de Geyn K. The influence of lung density corrections on treatment planning for primary breast cancer. Int J Radiat Oncol Biol Phys 14:179-190;1988.

— *Chapter 16* —

CODING & REIMBURSEMENT FOR CONFORMAL THERAPY

Jennifer O'Malley, MBA

As the use of three dimensional simulation and conformal therapy increases, radiation oncology practices have struggled to ensure that appropriate coding is used to bill for the services rendered. In response to the demand, in 1994 the Health Care Finance Administration (HCFA) introduced two new codes for the planning and treatment process for conformal therapy. Listed below are the descriptions from the 1994 CPT-4 book of these codes:

77295 — Therapeutic radiology simulation-aided field setting by 3D reconstruction of tumor volume.

77419 — Weekly radiology therapy management; conformal. Conformal is defined as multiple custom megavoltage treatment beams focused on a large 3D reconstructed target.

Since publication of these codes, there has been an animated debate as to the proper use of the technology and the coding. Since the technology for CT simulation and 3D treatment planning is not at the same level throughout the country, many practices have chosen to compensate for less than optimal systems with multiple verification simulations. These practices have made a decision to bill the patient for the additional simulations needed to verify information from their CT simulation and treatment planning systems. As a practice we feel that it is inappropriate to bill a patient for additional simulations due to the lack of a fully integrated 3D simulation and treatment planning system.

CODING POLICY

We have developed the following guidelines for coding 3D simulations, the associated treatment planning and weekly management of conformal therapy.

Three-Dimensional Simulation (77295)

To bill for this code, a patient must undergo a CT simulation and 3D reconstruction of the tumor volume.

If the CT simulation is utilized for localization purposes only (i.e., no DRRs or graphical planning is completed) the patient is billed for a complex simulation.

If a patient undergoes a CT simulation with 3D reconstruction and then has a simulation completed to verify bony landmarks, the patient is not billed for an additional simulation.

Weekly Radiology Management — Conformal (77419)

Any patients who have undergone a CT simulation and subsequent 3D reconstruction and graphical planning are highlighted throughout treatment as conformal patients and the weekly management is billed under 77419.

Treatment Planning (77263)

Each patient who undergoes a CT simulation and subsequent graphical planning is billed for a complex treatment plan.

Isodose Distributions — Complex (77315)

Billing for isodose distributions is completed on a per site basis. If additional graphical plans are completed for a cone-down, additional isodose distributions are billed. If the same plan is utilized for the cone-down, the patient is not billed again.

Medical Physics Consultation (77370)

If a physician requests a special consultation with the physicist during the planning process for a particularly complicated case, the physicist documents the consultation and the patient is billed for this code.

CODING AND BILLING PROCEDURE

Based upon the previously mentioned guidelines, the technical staff in the department completes either a manual charge ticket or enters the charges into a computerized charge capture system. Simulation and treatment planning coding are completed by the chief therapist and CT simulation therapist. Isodose distribution coding is completed by the dosimetrist assigned to complete isodose distributions for conformal therapy. Medical physics consultations are completed by the assigned physicist and are coded by that physicist.

All charges are accumulated on a daily basis, collated and transferred to the institution's technical and professional billing departments. These departments are responsible for entering charges into the billing system, distribution of bills to appropriate insurance carriers, and monitoring collection. The department maintains records of codes accumulated by the department and conducts quarterly audits to ensure that all charges captured by the department are subsequently billed to patients and their insurance carriers.

MEDICARE REIMBURSEMENT

Reimbursement for codes associated with conformal therapy appropriately reflects the amount of effort required to complete the process of 3D simulation and conformal therapy planning. Listed below are the global, technical, and professional component Medicare reimbursement for our geographical region. Please note that Medicare reimbursement may vary slightly depending upon your geographical location.

Table.

CPT Code	Description	Global	Professional	Technical
77295	Three dimensional simulation	1,200	139	968
77419	Weekly mgt. — conformal		205	
77263	Treatment planning — complex		178	
77315	Isodose plan — complex	173	89	92
77370	Medical physics consultation	493	385	119

Appendix I

SET UP UNCERTAINTY: SELECTED REFERENCES

Pelvis (Gynecological/Genitourinary)

1. Rabinowitz I, Broomberg J, Goitein M, McCarthy K, Leong J. Accuracy of radiation field alignment in clinical practice. Int J Radiat Oncol Biol Phys 1985;11:1857-1867.

2. Soffen EM, Hanks GE, Hwang CC, Chu JCH. Conformal static field therapy for low volume low grade prostate cancer with rigid immobilization. Int J Radiat Oncol Biol Phys 1990;20:141-146.

3. Ten Haken RK, Forman JD, Heimburger DK, Gerhardsson A, McShan DL, Perez-Tamayo C, Schoeppel SL, Lichter AS. Treatment planning issues related to prostate movement in response to differential filling of the rectum and bladder. Int J Radiat Oncol Biol Phys 1991;20:1317-1324.

4. Balter JM, Chen GT, Pelizzari CA, Krishnaswamy S, Rubin S, Vijayakuman S. Online repositioning during treatment of the prostate: a study of potential limits and gains. Int J Radiat Oncol Biol Phys 1993;27:137-143.

5. De Neve W, Van den Heuvel F, Coghe M, Verellen D, De Beukeleer M, Roelstraete A, de Roover P, Thon L, Storme G. Interactive use of online portal imaging in pelvic radiation. Int J Radiat Oncol Biol Phys 1993;25:517-524.

6. Rosenthal SA, Roach M, Goldsmith BJ, Doggett ED, Pickett B, Yuo HS, Soffen EM, Stern RL, Ryu JK. Immobilization improves the reproducibility of patient positioning during six-field conformal radiation therapy for prostate carcinoma. Int J Radiat Oncol Biol Phys 1993;27:921-926.

7. Schild SE, Casale HE, Bellefontaine LP. Movement of the prostate due to rectal and bladder distension: implications for radiotherapy. Medical Dosimetry 1993;18:13-15.

8. Kortmann RD, Hess CD, Jany R, Meisner C, Bamberg M. Reproducibility of field alignment in difficult patient positioning. Int J Radiat Oncol Biol Phys 1994;29:869-872.

9. Hunt MA, Schultheiss TE, Desobry GE, Hakki M, Hanks GE. An evaluation of setup uncertainties for patients treated to pelvic sites. Int J Radiat Oncol Biol Phys 1995. In press.

Head and Neck

1. Marks JE, Haus AG. The effect of immobilization on localization error in the radiotherapy of head and neck cancer. Clin Radiol 1976;27:175.

2. Rabinowitz I, Broomberg J, Goitein M, McCarthy K, Leong J. Accuracy of radiation field alignment in clinical practice. Int J Radiat Oncol Biol Phys 1985;11:1857-1867.

3. Huizenga H, Levendag PC, De Porre PMZR, Visser AG. Accuracy in radiation field alignment in head and neck cancer: a prospective study. Radiotherapy and Oncology 1988;11:181-187.

4. Halverson KJ, Leung TC, Pellet JB, Gerber RJ, Weinhous MS, Wong JW. Study of treatment variation in the radiotherapy of head and neck tumors using a fiber-optic radiotherapy imaging system. Int J Radiat Oncol Biol Phys 1991;21:1327-1336.

5. Rosenthal SA, Galvin JM, Goldwein JW, Smith AR, Blitzer PH. Improved methods for determination of variability in patient positioning for radiation therapy using simulation and serial portal film measurements. Int J Radiat Oncol Biol Phys 1992;23:621-625.

6. Dunscombe PB, Fox K, Loose S, Leszczynski K. The investigation and rectification of field placement errors in the delivery of complex head and neck fields. Int J Radiat Oncol Biol Phys 1993;26:155-161.

7. Hunt MA, Kutcher GJ, Burman C, Fass D, Harrison L, Leibel S, Fuks F. The effect of setup uncertainties on the treatment of nasopharynx cancer. Int J Radiat Oncol Biol Phys 1993;27:437-447.

Brain

1. Rabinowitz I, Broomberg J, Goitein M, McCarthy K, Leong J. Accuracy of radiation field alignment in clinical practice. Int J Radiat Oncol Biol Phys 1985;11:1857-1867.

See also Head and Neck, above.

Thorax

1. Marks JE, Haus AG, Sutton HG, Griem ML. Localization error in the radiotherapy of Hodgkin's disease and malignant lymphoma with extended mantle fields. Cancer 1974;34:83-90.

2. Marks JE, Haus AG, Sutton HG, Griem ML. The value of frequent treatment verification films in reducing localization error in the irradiation of complex fields. Cancer 1976;73:2755-2761.

3. Rabinowitz I, Broomberg J, Goitein M, McCarthy K, Leong J. Accuracy of radiation field alignment in clinical practice. Int J Radiat Oncol Biol Phys 1985;11:1857-1867.

4. Ross CS, Hussey DH, Pennington EC, Stanford W, Doornbos JF. Analysis of movement of intrathoracic neoplasms using ultrafast computerized tomography. Int J Radiat Oncol Biol Phys 1990;18:671-677.

5. Creutzberg CL, Visser AG, DePorre PMZR, Meerwaldt JH, Althof VGM, Levendag PC. Accuracy of patient positioning in mantle field irradiation. Radiother Oncol 1992;23:257-264.

Breast

1. van Tienhove G, Lanson JH, Crabeels D, Heukelom S, Mijnheer. Accuracy in tangential breast treatment setup: a portal imaging study. Radiother Oncol 1991;22: 317-322.

2. Creutzberg CL, Althof VG, Huizenga H, Visser AG, Levendag PC. Quality assurance using portal imaging: The accuracy of patient positioning in irradiation of breast cancer. Int J Radiat Oncol Biol Phys 1993;25:529-539.

Abdomen

1. Rabinowitz I, Broomberg J, Goitein M, McCarthy K, Leong J. Accuracy of radiation field alignment in clinical practice. Int J Radiat Oncol Biol Phys 1985;11:1857-1867.

General Issues Including the Incorporation of Set Up Uncertainties Into Treatment Planning

1. Goitein M, Busse J. Immobilization error: some theoretical considerations. Radiology 1975;117:407-412.

2. Goitein M. Calculation of the uncertainty in the dose delivered during radiation therapy. Med Phys 1985;12:608-612.

3. Urie MM, Goitein M, Doppke K, Kutcher GJ, LoSasso T, Mohan R, Munzenrider JE, Sontag M, Wong JW. The role of uncetainty analysis in treatment planning. Int J Radiat Oncol Biol Phys 1991;21:91-107.

4. Hunt MA, Schultheiss TE, Desobry GE, Hanks GE. Convolving setup uncertainties with dose distributions. Med Phys 1993;20:929. Abstract.

5. Bel A, Lebesque JV. Incorporation of random uncertainties into three-dimensional treatment planning. Med Phys 1994;21:914. Abstract.

Appendix II

TREATMENT PLANNING AND CT SIMULATION FORMS AND PROCEDURES

Fox Chase Cancer Center, Department of Radiation Oncology
Treatment Planning Information Sheet

Patient Name: _____   Simulation Date: _____

Physician: _____   Plan Due Date: _____

Treatment Plan

    Site: _____   Protocol: _____

    Description of Plan: _____

    Initial Fields: _____

    Cone Downs: _____

Prescription Doses

    Initial Fields: _____

    Cone Downs: _____

Critical Structures and Acceptable Doses

    Structure   Dose

    _____   _____

    _____   _____

    _____   _____

Suggested Beam Arrangements _____

_____

Previous Pertinent Irradiation _____

_____

Information Taken By _____

Information Given By _____

Fox Chase Cancer Center, Department of Radiation Oncology
CT Simulation Planning Worksheet

Patient Information and Scheduling

Name: _____   Date: _____

Treatment Site: _____   Physician/Resident: _____

Date of CT Sim: _____   Date of Conv. Sim: _____

Immobilization and Setup Instructions

Patient Positioning: _____

Immobilization Devices: _____

Setup Instructions: _____

_____

Scanning Information

Superior Scanning Limit: _____

Inferior Scanning Limit: _____

Slice Spacing: _____ Slice Thickness: _____

Contrast: _____

CT Sim Information

Isocenter Coordinates:

    Site: _____   Site: _____

    X: _____   X: _____

    Y: _____   Y: _____

    Z: _____   Z: _____

    Table Height: _____   Table Height: _____

Information Given By _____

Information Taken By _____

Index

Absorbed dose, 7
Acceptance testing, 6-22
ACQSIM, 27, 34, 36f, 43, 49, 152
Acquisition time, 53
Alpha Cradle, 107, 124, 129, 139, 164, 176
Aperture, 10, 25, 32, 36, 51, 52, 53, 126, 150, 167, 188
Aquaplast, 76, 89, 96, 152, 156t, 188
Attenuation coefficient, 41
Autocontouring, 27
Axial CT
    Data, 5, 9, 14, 47, 88, 92, 96, 100
    Scanner, 55, 57
    Slice thickness, 39, 47, 48, 52
Back projection algorithm, 5
Beam portals, 32
Beam's eye view (BEV), 18, 39, 47, 57, 58, 126, 193
Bite block, 76, 89, 96
Bolus, 91, 97, 173, 176
Boost field, 187, 188, 191, 191f
Brachytherapy, 135, 141, 142
Breast board, 188
Cerrobend blocks, 128, 132, 167
Clinical target volume (CTV), 31, 32, 34, 63, 64, 65, 66, 71, 76, 77, 78, 80, 81, 82, 89, 90, 91, 92, 96, 98, 101, 125, 126, 128, 129f, 131, 141, 142, 143f, 145f, 146f, 153, 154, 155t, 156, 157, 189, 189f, 190, 191f, 192f, 193f
Compensators, 34, 35t, 36
Complex geometry, 33
Compton scatter, 42, 45
Compton-weighted images, 42, 45, 46f
Contrast
    Agents, 26-27, 53, 72, 76, 80, 89-91, 96-97, 107, 108, 122, 142, 151, 155, 156t, 164, 165
    Setting, 27t, 28t, 35t, 36, 44, 45, 47, 48
Conventional simulation, 5, 18, 25, 26, 27, 30t, 32, 33f, 34, 36, 37, 38, 39, 47, 51, 53t, 57, 62, 65, 66, 67, 108, 109, 111, 112, 127, 128, 157, 164, 179, 187, 192, 194f
CT
    Data set, 1, 5, 9, 10, 12, 14, 16, 25-38, 40, 41, 42, 43, 54, 55, 66, 67
    Portogram, 111
    Scanner, 1, 5, 6, 6f, 7, 8, 10, 12, 22, 26, 32, 30t, 38, 39, 51, 52, 53, 54, 57, 55, 56t, 62, 65, 97, 106, 188
    Simulation, 5-7, 9, 12, 25-38, 51-58, 62, 63, 65, 66, 67, 71, 76, 78, 80, 81, 83, 89, 91, 92, 96, 98, 101, 103, 107, 108, 109, 110, 111, 112, 135, 137, 139, 153, 156, 157, 164, 165, 169, 170, 173, 176, 178, 179, 187, 189, 190, 191, 197, 198
    Treatment planning, 1, 61-68, 71, 78, 81, 83, 85, 91-92, 98-99, 101, 107, 108, 112, 120, 127, 129, 135, 149, 150, 152, 153, 156, 157, 164, 168, 179
Custom blocks, 36
Cyclical redundancy checks, 9
Data storage, 26, 55, 56, 56t, 57, 58, 58t
Diagnostic x-ray, 39
Digital composite reconstruction, 49
Digital rectal examination, 163, 164, 168, 170
Digitally reconstructed fluoroscopy, 42
Digitally reconstructed radiographs (DRRs), 13, 16, 25, 26, 27, 32, 33, 33f, 34, 34f, 36, 37, 39-49, 51, 52, 53f, 54, 55, 56, 57, 58, 58t, 62, 66, 67, 78, 80, 81, 82f, 83, 91, 92, 92f, 93f, 97, 98, 99f, 100f, 101, 109, 110, 111, 112, 126, 127, 128, 129, 130, 142, 155, 156, 157, 164, 169, 171, 177, 179, 180, 187, 191, 193, 197
Disk swapping, 56
Dose calculations, 27, 36, 51, 56, 62
Dose
    Limiting normal structures, 31, 33, 64, 65, 107, 117, 173
    Optimization, 36, 65, 127
    Volume histogram, 26, 79
    -Time parameters, 127
Dysphagia, 92, 123
Electron beam, 91, 187, 189, 190, 191, 191f, 192, 192f
Electron density, 7, 56
Electronic portal imaging devices, 47, 61, 67
Ethernet network, 54, 58
External beam radiation therapy, 149, 163, 164, 169, 170
Fan beam projection, 5, 12
Fast Ethernet, 54, 58
Field
    Alignment, 12, 34
    Aperture, 20, 32, 33f, 34f, 35t, 36, 66, 68

Design, 25-38, 61, 62, 65, 66, 67, 68, 92
Modification, 34
Size effect, 7
File server, 58
Film magnification, 19
Fluoroscopy, 25, 43, 51, 57
Gantry mounted laser, 10, 12
Geometric optimization, 33, 34, 36, 127
Graphical isodose planning, 65, 66-68
Gray scale mapping, 40
Gross tumor volume (GTV), 27, 31, 63, 65, 66, 76, 76t, 77, 80, 81, 81f, 82f, 89, 96, 107, 108, 125, 126, 129, 141, 142, 152t, 153, 166, 166f, 167t, 169, 171, 176, 177, 178, 178t, 189
Health Care Finance Administration, 197
Helmholtz coil, 164, 174, 176, 180
High contrast resolution, 7, 8
ICRU Report #50, 1, 31, 63-66
Image
  Artifacts, 27
  Fusion, 26, 62
  Intensifier, 5, 51, 157
  Parameters, 9, 26
  Quality, 5, 9, 44, 45, 46, 49, 62, 78, 155, 158
  Reconstruction, 12, 26
Imaging protocols, 26
Immobilization, 28t, 31, 61, 63, 64-65, 76, 96, 107, 124, 128, 139, 152, 156, 156t, 157, 164, 167, 176, 188
Intelligent contouring, 27
Interpolation routines, 17
Interscan delay time, 53
Intraoperative radiation therapy, 180
Iso-CT mapping, 49
Isocenter
  Calculation, 10, 14
  Marking, 12, 54, 56
  Movement, 16
  Positions, 10, 19, 34, 56
Laser patient marking system, 6, 9-10, 62
Lateral
  Isocenter position, 10, 56, 156
  Opposed beams, 156, 157
Line integral algorithms, 41, 43
Linear accelerator, 47, 156
Linear array of detectors, 5
Linear attenuation coefficient, 41
Local area network (LAN), 9, 34, 67, 168
Localization, 5, 14, 18, 25-38, 47, 51, 54, 59, 61, 62, 63, 64-67, 76, 78, 99, 90, 91, 92, 96, 97, 98, 107-108, 110, 111, 112, 124, 126, 129, 141, 150, 152, 154, 156, 164, 165, 176, 165, 176, 178, 189, 190, 197
Look-up tables (LUTS), 42, 44, 45, 47
Low contrast resolution, 7, 8
mAs linearity, 7, 8
Measurement tolerances, 9, 13
Medicare reimbursement, 198
Modulation transfer function, 14
MRI, 27, 89, 95, 96, 106, 123, 150, 152, 153, 155t, 164, 165, 168, 170, 176, 180
Oblique angles, 30f, 33, 36, 124, 132, 154, 156, 159f
Opposed anterioposterior fields, 32, 127, 128
Organ motion, 31, 64, 96, 107, 167
Partial brain irradiation (PBI), 150, 151, 155, 157
Partial volume averaging, 47
Patient coordinate system, 25, 31, 36, 65, 66, 67, 68
Patient positioning, 28t, 31, 32, 65, 76, 89, 96, 126, 128, 152, 153, 156, 164, 188
PET, 27
Photoelectric scatter, 42, 44, 45
Photoelectric-weighted images, 42, 44, 46f
Picture archiving and communication system, 47
Pilot view, 27, 30t, 31f, 32
Pixel size, 16, 19, 40, 45
Plan evaluation, 36
Planning target volume (PTV), 31, 32, 33f, 34f, 63, 64, 65, 66, 76f, 77, 78, 78t, 79, 80, 80f, 81, 81f, 82f, 90, 91, 92t, 96, 98, 99t, 107, 108, 110, 110f, 111, 112f, 125, 126, 127, 128, 129, 130, 130f, 131, 131f, 132, 141f, 144f, 146f, 153, 154t, 167, 167t, 169, 170, 170f, 171, 171f, 176, 178, 178f, 179, 180, 187f, 189
Portal field aperture, 6
Portal image, 2, 36, 47, 67, 106, 132, 164, 168, 171, 169f, 172f
Prescription dose, 64, 78, 79t, 108, 110, 126, 127, 168, 179
Quality assurance, 5-18
Radionuclide bone scan, 164
Radiosurgery, 47
Ray line divergence, 14, 15f, 16, 16t, 17, 17f, 40, 47
Ray tracing, 17, 40, 41, 42, 43, 44, 49, 56
Real time portal imaging, 168

Registration reconstruction error, 14
Room shielding requirements, 7
Room's eye view, 34, 35t
Rotation matrix, 19
Scan reconstruction circle, 27
Set up
    Accuracy, 46
    Variation, 64, 107, 141, 155t
Simulation film, 16, 36, 37, 45, 62, 67, 92, 128, 135
Simulator with a CT option (sim CT), 5
Single modality adjuvant therapy, 173
Slice thickness and spacing, 7, 1, 19, 21, 26, 28t, 47, 48f, 52, 63, 78, 91, 92, 155, 156t, 164
Soft tissue visualization, 42, 45, 47, 49, 128
Solid shading, 45
Source-to-film distance, 40
Source-to-object distance, 52
Spatial
    Linearity, 10, 11, 12, 12t, 14t, 17, 18, 19t, 20t
    Resolution, 7, 37, 66
    Uniformity, 7
SPECT, 27
Spiral CT scanners, 6, 26, 53-54, 55, 55t, 56t, 62, 106
Standard beam arrangement, 34
Stereotaxic irradiation, 149, 150
Structure contours, 36
Successive refinement, 49
Summation, 40, 40f
Surface markers, 27
Tape backup, 55, 58
Target
    Contour verification, 14, 20t
    Localization, 13, 14, 17, 20t, 165
    Volume geometry, 26
    Volume localization, 2
Technique chart, 36
Test phantoms, 7, 10, 14, 15
Therapy alignment gauge, 10
Three dimensional
    Conformal radiation therapy, 116, 124, 129, 132, 163, 197, 198
    Contouring, 27
    Simulation, 197-198
    Visualization, 41, 49
    Volume rendering, 41
Tolerance dose, 64, 78, 107, 129, 141, 155, 168, 177

Transaxial resolution, 53, 54, 55
Treatment
    Field, 25, 26, 34, 35t, 36, 37, 51, 53f, 54, 61, 63, 65, 66, 67, 108, 110, 126, 127, 128, 135, 140, 142, 189, 192f, 194f
    Intent, 61, 62-63
    Planning system (TPS), 1, 25, 33, 36, 39, 49, 52f, 62, 63, 65, 67, 68, 127, 152
    Verification, 25, 47, 48, 61, 62, 78, 98, 109, 128, 132, 142, 156, 157, 168, 179
Triangulation points, 32
Trilinear interpolation, 41, 42, 43, 44
Tube loading, 54
Tumor volume isocenter, 6
Verification images, 25, 36-37
Vertex fields, 29t, 34
Virtual simulation, 6, 7, 12, 13, 14, 17, 18, 18t, 19, 20t, 25, 27, 30t, 32, 33, 33f, 34, 35t, 36f, 37, 39, 51, 52f, 53f, 54, 56, 57, 58, 67, 68, 78, 90, 91, 97, 98, 126, 155, 157, 178, 189, 190
Virtual simulator, 6, 9, 13-14, 25, 33, 34, 36, 49, 56, 57, 58, 58t, 62, 65, 66, 67, 68, 127, 156
Volume averaging effects, 26, 47, 108, 155
Volume element (voxel), 40, 40f, 41, 42, 43, 44, 47
Volume of interest (VOI), 43, 44, 45, 49
Volume rendering, 41, 49
Water phantom, 7
Wedge-pair beams, 153, 156
Wedges, 128, 132
Whole brain irradiation (WBI), 150, 151, 152
Wireframe shading, 45